Liberation Science: Putting Science to Work for Social and Environmental Justice

Edited by Steven H. Emerman[1], Marcia Bjørnerud[2], Jill S. Schneiderman[3], and Sarah A. Levy[4]

[1]Utah Valley University, Orem, Utah

[2]Lawrence University, Appleton, Wisconsin

[3]Vassar College, Poughkeepsie, New York

[4]U.S. Forest Service, Portland, Oregon

ISBN: 978-1-300-43792-5

Table of Contents

Part I: Redefining Ecosystems: Mapping the Hidden Ecologies of Environmental Injustice

Part II: Recruiting Citizen Scientists: Inclusion and Exclusion in Community-Level Environmental Activism

Part III: Rediscovering the Commons: Philosophical Principles for Global Environmental Justice

Introduction: What is Liberation Science?

Steven H. Emerman and Marcia Bjørnerud

In the four decades since the first Moon landing, astonishing advances in digital technology have radically changed the fabric of our personal and social lives through electronic devices that the Apollo astronauts themselves could only dream about. Biotechnology and medical research have similarly yielded almost miraculous results: unprecedented agricultural productivity and new ways to defy death (at least temporarily). If we measure those same four decades as the time since the first Earth Day, however, we see almost no progress in some far more mundane matters, including public vulnerability—especially in poor and minority communities—to environmental health risks and natural hazards. Why, in an era of once-unimaginable technological achievement, are the poorest citizens of the world's richest country still sickened by the air and soils in their neighborhoods? Why, at a time when we have sent robotic rovers to find water on Mars, is clean, fresh water increasingly inaccessible to millions of humans back on Earth?

There are many explanations and excuses for the asymmetry between our dizzying progress in high technology and our abysmal failure to make progress on issues of environmental health. Environmental injustice is of course entangled with deeply rooted social structures and political habits that are not easily changed. But there is another, equally important reason that these problems continue to seem so intractable: the resources and tools of the scientific community have simply not been applied to social and environmental injustice in the same systematic, single-minded way that made space exploration, high-speed computing and genome sequencing possible.

Many scientists are understandably wary of engaging themselves in research that is too closely connected with political issues. Science must maintain its capacity for objective inquiry, or it risks devolving from a powerful illuminator of the unknown into an instrument of dogma and censorship. We were taught in school that one of the beauties of science is the fact that all of us—rich and poor, capitalists and communists—share the exact same physical laws. One of us was told by a professor of mineralogy that if he ever met an alien from another planet he could say with confidence, "Thirty-two crystal classes!" and the alien would reply "Same here!" But it is really a trivial point that we all share the same physical laws. Few scientists are actively engaged in the practice of discovering new physical laws. The real question is: In whose interests, for the solutions of whose problems, are the existing, recognized physical laws being put into practice?

Few practicing scientists give much consideration to the pervasively political framework within which they carry out their work. Budget allocations from government funding agencies, hierarchies within and among scientific societies, and the nature and prestige of various scientific prizes all reflect political priorities. These priorities, in turn, inevitably shape the course of scientific progress because young scientists are naturally drawn to fields that offer ample funding and opportunities to achieve recognition and honors. Fields flush with money and the cachet of glamour therefore thrive, while other possible lines of inquiry simply never develop.

It is time to acknowledge that continuing failure to turn our formidable scientific prowess toward issues of environmental justice is itself a political decision. Under the banner of American patriotism, we have chosen to pursue science driven largely by military and commercial interests, believing that benefits eventually spin off or trickle down to ordinary people. The defense and aeronautics industries have indeed provided us with innovations ranging from space food sticks to global positioning systems. But one wonders how much more could have been achieved if scientific funding decisions had placed equal weight on upholding constitutional guarantees of civil rights and equal protection.

The above description of a social-scientific system in which scientific talent and energy inevitably promote the interests of those with power and money seems to imply a hopeless situation for the poor who are disproportionately affected by environmental degradation. In fact, all deep-seated social problems seem intractable without a careful consideration of two of the most important words in social science, which are ideology and hegemony. Ideology refers to the set of beliefs that justify the power and perpetuation of the ruling class. (Although reasonable people may disagree as to who is truly part of the ruling class, the alternative concept that all of us in our society share equal access to power and resources is simply absurd.) The dominant ideology of modern Western civilization is the intersection of nationalism and capitalism. For the purpose of understanding why scientific energy is not placed at the disposal of solving the environmental problems faced by the poor, the key aspects of the dominant ideology are:

1) The natural order of the world is the division of people into nations. Although there are some mutual obligations among the citizens of a nation, what a citizen of one nation owes to a citizen of another nation is extremely limited.
2) The natural order of the world is that the obligations of one person toward another are based upon the other's ability to pay or otherwise offer something in exchange. Even among the citizens of a nation, what the rich owe to the poor in the absence of adequate payment is extremely limited.
3) Social and environmental problems disproportionately faced by the poor may have existed in the past, but they have already been solved.

The second key word, hegemony refers to the process by which all people, even or especially those who are not members of the ruling class, are led to believe in the dominant ideology. Hegemony is the process by which people are led to accept a set of beliefs that acts against their interests. Hegemony is accomplished by parents, the media, churches, and especially by educational institutions. The key aspect of any ideology is that there are no alternatives to the current socio-economic system. The current socio-economic system is the natural way, the only way that the world can be organized. Our task as academics, as scientists, as activists, as human beings, is not simply to accept and act within the current state of affairs, but to imagine alternatives and then to make those alternatives happen. Any time one cannot imagine alternatives, one should regard oneself as a victim of hegemony. One of us once asked the students in an Introduction to Environmental Studies class whether they could imagine that one day, the U.S. Government would be taken over by environmental activists who would reform American society under ecologically sustainable lines. Only one hand went up. The students were not being asked to make a contribution, give up their education, quit their jobs, or assassinate somebody, but simply to use their imagination. This is the power of hegemony.

The purpose of this volume of essays is to imagine alternatives, to provide a variety of perspectives on the role of science in the movements for social and environmental justice, acknowledging its past culpability for creating and ignoring problems, and exploring its future capacity to contribute to the solutions. Through a diverse set of case studies, the authors—a diverse group of natural and social scientists, many of them young scholars of color— provide examples of how the analytical approaches and conceptual frameworks of many different scientific disciplines can help document, mitigate, and resolve instances of social and environmental injustice. The essays illustrate that no single methodology will ever be able to address these problems, which are arguably more complex than those of particle physics or molecular biology because they cannot be addressed with a reductionist mindset. Instead, this emerging science for social and environmental justice will require broadly trained scientists who are able to think multidimensionally as well as multiculturally, who interact with the communities they study, and who are able to visualize new educational and economic systems that foster both human and environmental health.

In the 1960s a courageous group of Catholic clergy and laypeople created Liberation Theology, a religious practice that intentionally took the side of the poor—despite the fact that the rich and poor have the same God, just like they have the same laws of nature. Liberation Science is the practice of intentionally using the knowledge and methods of science to solve the social and environmental problems faced by the poor. Liberation Science is science that is able to solve the social and environmental problems faced by the poor because it has been freed from the ideology of nationalism-capitalism.

This book is organized around three central ways that science can be put to work for the cause of social and environmental justice:

1) Scientists need to broaden the definition of ecosystems to include people, cultures, and urban landscapes if we are to understand environmental injustice; the scientific preference for studying "pristine" ecosystems is itself a cause of environmental injustice. Not only must ecosystems become more inclusive, but the possibility that an ecosystem, or even the whole Earth, may, in some ways, act as a single organism must be considered.
2) The logic and methods of science must be made available to ordinary people, empowering them to understand the ecologies of their own communities.

3) In addressing environmental problems, science should be open to complementary philosophical approaches that draw upon cultural and spiritual traditions of particular regions or communities.

Each theme corresponds to a part of the book. Part I of the book, "Redefining Ecosystems: Mapping the Hidden Ecologies of Environmental Injustice" includes five chapters that illustrate how redressing environmental injustices requires that the ecological sciences adopt both more expansive and more holistic definitions of ecosystems to include humans, cultural practices, and the built environment, together with the possibility that an ecosystem could mimic the behavior of a single organism. In "Poisoned Minds: The Legacy of Lead Poisoning among Urban Youth," Gabriel Filippelli, Mark Laidlaw and Deborah Morrison report how, contrary to widespread popular belief, the urban poor continue to face unacceptably high levels of exposure to ambient lead even after more than a half century of efforts to minimize its use. They show that understanding patterns and pathways of lead exposure can make it possible to design simple strategies to mitigate lead poisoning. Filippelli and co-workers further argue that scientists have a responsibility not merely to document potential threats to human health but to see that their work is translated into practical and effective policy. This call to action is echoed by Victoria Kalkirtz, Michelle Martinez and Alexandra Teague in their chapter, "Fishing on the Detroit River: Historical Pollution, Environmental Racism, and the Scientific State," in which they provide evidence for large disparities between racial and economic groups in awareness of potential hazards from eating fish from a polluted urban river. In their chapter "Mothers Shouldn't have to be Scientists," John Dao and Rebecca Roberts explain why the poor bear a disproportionate exposure to the effects of the widespread toxic chemical bisphenol A. They discuss further how the chemical industry and regulatory agencies have prevented poor mothers and others from receiving the information necessary to make informed consumer and lifestyle choices that would affect exposure to bisphenol A.

Sarah Levy's chapter "'We Need More Than Thirty Trees': A Case for Urban Forestry" reveals that city foresters face difficulties not only for all the expected reasons—poor air and soil quality, limited sunlight and space—but also because of prejudices within the academic culture of forestry, which tends to dismiss urban settings as "unnatural" or "degenerate." In light of the many social benefits of healthy urban forests, Levy calls for a reexamination of the assumptions and priorities of forest science. In the final chapter in the first part, "Self-organizing Systems and Environmental Justice: Application to Arsenic Contamination of Groundwater in Nepal," Steven Emerman, Aimee Luhrs, Susan Sandford, and Adam Finken contest the dominant scientific paradigm that the arsenic contamination of groundwater in Nepal is unrelated to any human activity and argue that the flawed scientific paradigm is connected with a flawed social paradigm that does not recognize the rights of the poor. They show that the recognition that an ecosystem, in some ways, mimics the behavior of a single organism, leads to the conclusion that the conservation and restoration of healthy grasslands and forests could control the arsenic contamination of groundwater.

The second part of the book, "Recruiting Citizen Scientists: Inclusion and Exclusion in Community-Level Environmental Activism," provides examples of grassroots efforts to raise environmental awareness at the community level. The two chapters describe how sharing the tools of science with ordinary citizens can catalyze action on environmental issues. In her chapter "Public Participation and Spatial Information," Shalini Vajjhala explores the potential of community cartography as a tool for sharing scientific information and fostering engagement in environmental decision making. Arguing that "participation and inclusion are cornerstones of social justice," Vajjhala suggests that the act of making maps involves collective re-imagining of shared spaces and can be a powerful tool for catalyzing social change. In their chapter "Using School-Based Environmental Science Projects to Further Environmental Justice," Nicky Sheats, Theodore Carrington, Fletcher Harper, Kim Gaddy and Valorie Caffee describe the New Jersey Urban Air Quality Education and Awareness Initiative, an innovative program in which high school students monitor airborne particulate matter in their neighborhoods and also work with community leaders in developing policies to improve air quality. The provocative second chapter of this part, "Between a Rock and a Green Place: Exploring the Relationship between Green Consumerism and Social Justice," by anthropologist Tendai Chitewere, analyzes the rhetoric and reality of an upscale "ecovillage" in New York State, applauding its aspirations but positing that such communities, with their emphasis on "green" consumerism and de facto exclusion of low-income families, may actually perpetuate the culture of environmental injustice.

The final part, "Rediscovering the Commons: Philosophical Principles for Global Environmental Justice," includes two essays that propose radical (and radically different) changes in the paradigms for global environmental thinking. In "Buddhist Living in the Anthropocene," geologist Jill Schneiderman argues that the scientific practice of "divorcing head from heart" is unnatural and undesirable and that there should be a place in scientific discussions of environmental issues for the principles of ethical conduct, especially as articulated in Buddhist teachings. In "Geomimicry for Social and Environmental Justice," Marcia Bjørnerud, also a geologist, proposes that environmental wrongs and social injustices are both rooted in a distorted perception of the relationship between humans and the natural world. She suggests that Earth's own history provides both practical and philosophical lessons about building stable and durable systems and that our collective well-being depends on respecting and emulating geologic laws rather than trying to evade them.

The first step in liberation is imagining an alternative to the current order. If one reader reads this book and imagines his or her own alternative to the current practice of science, we will regard this volume as a success. On that basis, we welcome a dialogue with our readers. Any reader should feel free to contact the author of any chapter through Steven Emerman, the lead editor, at StevenE@uvu.edu. We leave the reader with one last question: What if every practice, every discipline, had a liberation version, a version that intentionally took the side of the poor? What if there were a Liberation Finance or a Liberation Auto Mechanics? What kind of world would that be?

About the Contributors

Marcia Bjornerud is Professor of Geology and Environmental Studies at Lawrence University in Appleton, WI. Dr. Bjornerud's research focuses on the physics of earthquakes and mountain building, and she combines field-based studies of bedrock geology with quantitative models of rock mechanics. She is the author of a popular book, <u>Reading the Rocks: The Autobiography of the Earth</u> (Perseus/Basic*),* and was a Fulbright Senior Scholar in Norway (2000-2001) and New Zealand (2009).

Tendai Chitewere is an Assistant Professor in the Liberal Studies Program at San Francisco State University. She is an anthropologist who conducts ethnographic research on sustainable communities, ecovillages, and urban agriculture in the United States; specifically she questions cultures of (green) consumption and advocates for lifestyle changes that strive for simplicity to achieve sustainability. Born and raised in Zimbabwe, Tendai grows some of her food in her North Oakland urban yard.

Dana Coelho works with the U.S. Forest Service in Golden, CO. She has Masters Degrees in Sustainable Development & Conservation Biology and in Environmental Policy. Her passions include urban ecology and the blending of social and physical sciences for more sound long-term conservation practices.

John Dao is an undergraduate studying Biochemistry and Molecular Biology at Ursinus College in Collegeville, PA. Before attending Ursinus, he was a student researcher at Drexel University College of Medicine and has presented his research at numerous science fairs and conferences. In 2010 he was appointed to the Philadelphia Youth Commission, a part of city government that represents Philadelphia's youth in public hearings and gives testimony that reflects the youth perspective.

Steven H. Emerman is an Associate Professor in the Department of Earth Science at Utah Valley University and specializes in arsenic and heavy metals in surface water and groundwater. He was a Fulbright Professor at Tribhuvan University in Nepal in 2003 and is a frequent visitor to Nepal. From 1986-1989 he served in the military wing of the African National Congress in the areas of intelligence and psychological operations.

Gabriel M. Filippelli is a Professor in the Department of Earth Sciences at Indiana University - Purdue University Indianapolis (IUPUI) and specializes in urban environmental burdens to human health. He is the Director of the Center for Urban Health at IUPUI, is the Chair of the Science Planning Committee for the Integrated Ocean Drilling Program (IODP), and was the recent past Chair of the Geology and Health Division of the Geological Society of America. From 1987-1989 he was a U.S. Peace Corps Volunteer in the Republic of Kiribati, serving as a Vocational Skills Trainer on the outer island of Onotoa.

Adam Finken graduated from Simpson College in 2004 with a B.A. in Environmental Science and now works as a geotechnical engineering technician.

Victoria Kalkirtz has a Master of Science in Environmental Justice and a Master of Urban Planning with a focus on Environmental Planning and Land Use from the University of Michigan. She currently lives in Madison, WI, as a Project Coordinator for the IPM Institute, working to reduce the use of toxic pesticides in schools, public spaces and agriculture. Victoria's interest in environmental justice stems from growing up in Chicago and witnessing environmental and social inequities on a daily basis.

Mark A. S. Laidlaw is an Environmental Geoscientist affiliated with Macquarie University in Sydney, Australia, and has specialized in children's exposure to lead in soils and dusts since 1993. Mark has also been employed as a contaminated land scientist and coal exploration geologist.

Sarah A. Levy works with the U.S. Forest Service Legislative, Public Affairs, and Partnerships Office in Portland, OR. She received an M.S. from the University of Michigan, School of Natural Resources and Environment, with emphases on environmental justice and environmental policy. In her free time, Sarah enjoys good books, great food, and spending time exploring the beautiful Pacific Northwest.

Aimee J. Luhrs graduated from Simpson College in 2006 with a B.A. in Environmental Science and English and now works as a water pollution control specialist for the City of Indianola, Iowa.

Michelle Martinez has a Master of Science from University of Michigan School of Natural Resources and Environment in Environmental Policy and Planning, with a focus on Environmental Justice. She currently lives and works in Detroit as an environmental justice activist. In 2005 she studied community aqueduct management in the coffee growing region of Colombia, and broadened her interest in community resource management in her work for IFRI in 2007.

Deborah E. Morrison is a Ph.D. student in Applied Earth Sciences in the Department of Earth Sciences at Indianapolis University - Purdue University at Indianapolis with a specialization in environmental health, specifically lead poisoning in children. She has earned graduate degrees in earth sciences and epidemiology and had a two-year fellowship for urban educators where she took her graduate work into the classroom to allow inner city high school students the opportunity to participate in real-world lead contamination projects. Debbie lived in the Middle East for five years, then moved to South Florida to help operate a clinical chemistry reagent manufacturing company before moving back to Indiana to pursue higher education.

Paula Randler works with the U.S. Forest Service Cooperative Forestry office in Washington, D.C. She earned her Master's Degree in Environmental Management, focusing on the social science of conservation and development, including practical experience in urban forestry with local low-income residents. Paula's passions include social and environmental justice, urban poverty alleviation, and all kinds of process improvement.

Rebecca Roberts is an Associate Professor in the Department of Biology and Coordinator of the Biochemistry and Molecular Biology Program at Ursinus College in Collegeville, PA, where she studies the effect of hormonal regulation of the immune system, with a focus on the roles of estrogen and bisphenol A in Systemic Lupus Erythematosus. After completing a postdoctoral fellowship at Harvard Medical School, she became a Visiting Professor at Carleton College in Northfield, MN, where she first became interested in the environmental estrogen, bisphenol A. As a scientist and mother of three young children, she has been active in educating the public about bisphenol A and recently presented expert testimony to the Pennsylvania House of Representatives Consumer Affairs Committee regarding toxin-free toddler and baby products.

Susan E. Sandford is a graduate of Simpson College and holds a Bachelor of Arts degree in Environmental Science and a minor in Geology. From 2007 to 2010 she served as a Community-Based Environmental Management volunteer for Peace Corps Peru, working towards finding a balance in her community between development and conservation.

Jill S. Schneiderman is a Professor in the Department of Earth Science and Geography at Vassar College, where she also teaches in the multidisciplinary programs in Women's Studies and Environmental Studies. She was a Fulbright Scholar at the University of the West Indies in St. Augustine, Trinidad, in 2003, where she conducted research on women and water resources in the Caribbean. In 2009-2010 she was a Contemplative Practice Fellow of the Center for Contemplative Mind in Society, for whom she developed a curriculum in Earth Science that integrates contemplative practices into pedagogy.

Nicky Sheats is the director of the Center for the Urban Environment of the John S. Watson Institute for Public Policy at Thomas Edison State College, where he is extremely active in environmental justice issues through his work with the New Jersey Environmental Justice Alliance and other environmental justice organizations. Sheats was trained as a biological oceanographer by the Department of Earth and Planetary Sciences at Harvard University, from which he received his Ph.D., but has concentrated on air pollution issues since serving as a post-doctoral fellow at the Earth Institute of Columbia University. He began his professional career as a public interest attorney and worked as a legal services attorney and as a public defender after graduating from Harvard Law School.

Alexandria Teague graduated with a Master of Science in Natural Resources specializing in Environmental Justice and a Master of Urban Planning specializing in Environmental Planning and Land Use from the University of Michigan. As a planner she works to integrate environmental justice in the planning process, ranging from the development of land use policies to community outreach. Alexandria's dedication to protecting the natural and human environment from pollution was enhanced during her exchange program to Brazil in 2000, in which she was able to see sights like the Amazon Rainforest.

Shalini Vajjhala currently serves as Deputy Assistant Administrator for International and Tribal Affairs at the U.S. Environmental Protection Agency. Prior to joining the EPA, she was Deputy Associate Director for Energy and Climate at the White House Council on Environmental Quality and a Fellow at Resources for the Future, an environmental think tank in Washington D.C. Her academic research focuses on using innovative interdisciplinary mapping and spatial analysis tools and methods to evaluate public policy issues, including siting major energy facilities, regulating environmental justice, managing carbon sequestration, facilitating community-based resource management, and adapting to climate change.

Thanks to Reviewers

We are grateful to the following people who reviewed individual chapters of the book:

Virginia Ashby Sharpe, Veterans Administration, National Ethics Center

Caryn Bosson, TreePeople

Dan J. Brabander, Wellesley College

Bill Burch, Yale School of Forestry and Environmental Studies

Will Craig, University of Minnesota

Amity Doolittle, Yale School of Forestry and Environmental Studies

Josef Eckert, University of Washington

Cheryl Erler, Indiana University School of Nursing

Sally Fairfax, University of California - Berkeley

Mary Gray, Indiana University

CoryAnne Harrigan, Simpson College

Rick Jarow, Vassar College

Patrick L. Kinney, Columbia University

Lee Klinger, Sudden Oak Life

Kristen Kurland, Carnegie Mellon University

Joseph Nevins, Vassar College

Robert Nixon, University of Wisconsin

Vasantha Padmanabhan, University of Michigan

Geoffrey S. Plumlee, U.S. Geological Survey

Maria C. Powell, Madison Environmental Justice Organization and Nanotechnology Citizen Engagement Organization

Ron Sharp, Vassar College

Peggy Shepard, WEACT for Environmental Justice

Dan Stephen, Utah Valley University

Paul Tayler, Utah Valley University

Debbie Walser-Kuntz, Carleton College

Sacoby Wilson, University of South Carolina

Part I: Redefining Ecosystems: Mapping the Hidden Ecologies of Environmental Injustice

Chapter 1: Poisoned Minds: The Legacy of Lead Poisoning Among Urban Youth

Gabriel M. Filippelli, Mark A.S. Laidlaw and Deborah E. Morrison

Summary

While significant headway has been made over the past 50 years in understanding and reducing the sources and health risks of lead (Pb) and Pb poisoning, the incidence of Pb poisoning remains high in urban regions of the U.S. At particular risk are poor children of color who inhabit the polluted centers of our older cities without the benefits of adequate nutrition, education, and access to health care. To provide a future with fewer environmental and health burdens related to Pb, we need to adopt a completely different paradigm of the exposure pathway of children to Pb, namely the understanding that the cause of chronic Pb loading to urban youth is mostly their continued contact with dust derived from Pb-enriched inner city soils. This soil acted as a highly efficient absorber of human-emitted Pb over about 100 years of urban development, and is now returning that Pb to the next generations of people unfortunate enough to be members of the urban poor. This new paradigm, verified by our recent research in deciphering the causes of a seasonal oscillation in children's blood Pb levels, points to a relatively simple and cost-effective way forward in reducing the Pb load for urban youth.

Introduction

There is a common—but misguided—perception that lead (Pb) poisoning is no longer a threat. Indeed, effective regulations against leaded gasoline and lead-based paint have dramatically reduced Pb exposure. But as new research demonstrates, the threat to urban neighborhoods across the nation is still very real.

In the 20th century, two new applications turned lead toxicity into a widespread problem. First, Pb-based paints became the gold standard for new homes in the early part of the century, prized for their durability and bright white color. Second, Pb additives for gasoline were developed as an anti-knock engine formula in the 1920s, and the explosion of motor vehicles in the middle part of the century was fueled by gasoline doped with tetra-ethyl Pb. By the 1970s, Americans encountered Pb at every turn (Mielke 1994).

It took the efforts of Clair Patterson, a Cal Tech geochemist, to bring this hazard to the public's attention. In the 1950s, Patterson was conducting experiments designed to pinpoint the age of various rocks—but found that his results were skewed by consistent Pb contamination. Further studies showed that Pb levels were elevated in water, soil, even Arctic ice—and most troubling, in the human body. Over the next three decades, Patterson waged a lonely crusade against Pb that attracted the vociferous opposition of industry groups. But finally he convinced lawmakers and regulators to outlaw Pb in pipes, solder, and eventually in gasoline. As a result of Patterson's efforts, and those of a number of public health advocates, the number of children affected by Pb poisoning has been reduced by over 80% (Fig. 1.1). In the movies, Patterson's triumph would signal the closing credits—but in the real world, the story continues.

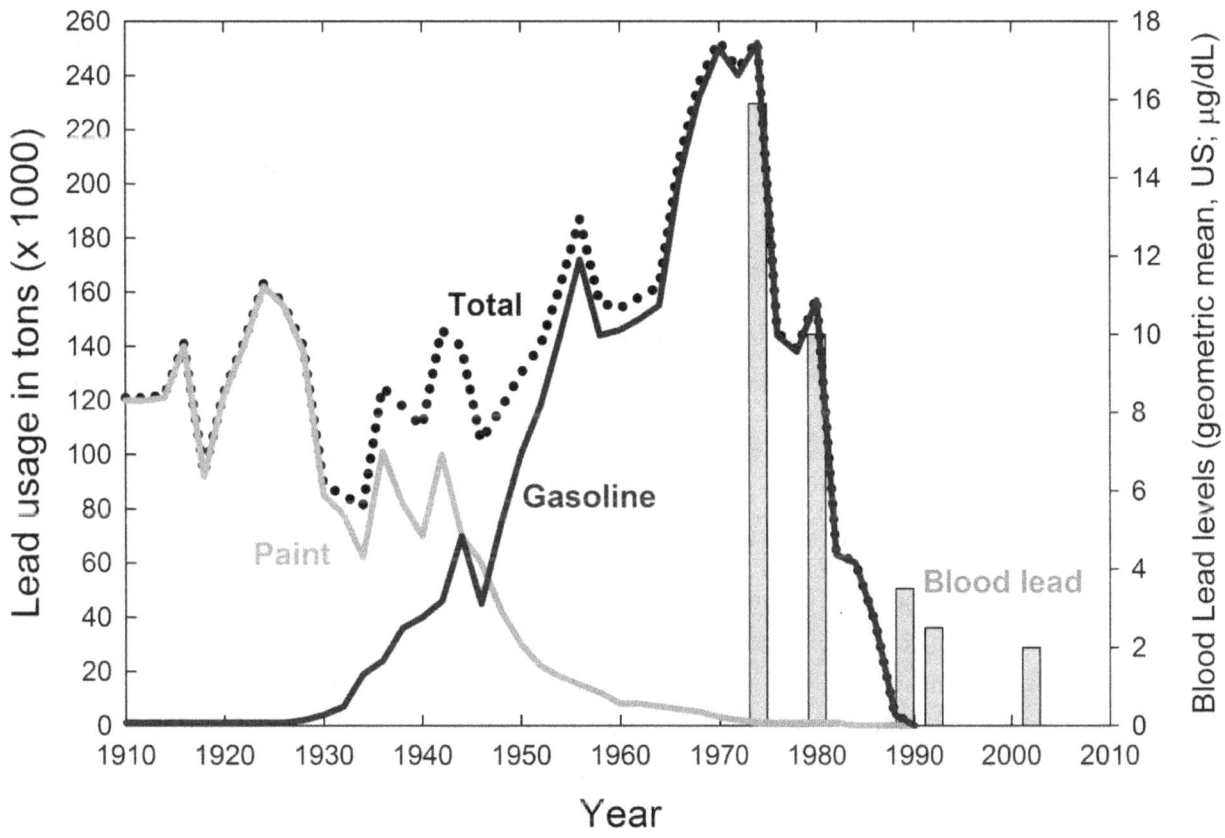

Fig. 1.1. History of Pb usage in paints and in gasoline (after Mielke (1999)), and US geometric mean blood lead levels (BLL) (after NHANES (2008)). Lead usage shows the early dominance of Pb-based paints followed by the boom in transportation resulting in a high use of leaded gasoline. The decline after the 1940s in Pb-based paints and after the 1970s in Pb from gasoline points to the environmental controls put in place on both of these industrial sources. The earliest US population subsample for Pb levels in blood occurred in the mid-1970s. In subsequent years, the mean BLL decreased along with the decrease in Pb usage, until sometime in the mid-1980s, when the decline in BLLs flattened, indicating continued sources of Pb from the legacy of industrial Pb usage. (Color version is available in the EBook or from the author at StevenE@uvu.edu.)

While less than 2% of children in the U.S. now suffer from Pb poisoning today (a value much improved but still a serious public health epidemic), children living in urban centers have Pb poisoning rates several times the national average. In 1980, Clair Patterson presaged the current state of environmental insults to urban populations: "Sometime in the near future it probably will be shown that the older urban areas of the United States have been rendered more or less uninhabitable by the millions of tons of poisonous industrial Pb residues that have accumulated

15

in cities during the past century" (NRC 1980). While many might consider Pb poisoning a closed chapter in the annals of public health, recent research shows that the dangers still exist, and are elevated among the most at-risk children in our society.

In this chapter, we discuss how children are exposed to Pb and how they are affected, where the continued sources of Pb are in urban areas, and how Earth scientists can assist health scientists in enhancing the health of the population—particularly the poor people of color who inhabit the polluted centers of our older cities without the benefits of adequate nutrition, education, and access to health care. Achieving this level of social justice involves adopting a completely different paradigm of the exposure pathway of children to Pb, namely the understanding that the cause of chronic Pb loading to urban youth is mostly their continued contact with dust derived from Pb-enriched inner city soils. This soil acted as a highly efficient absorber of human-emitted Pb over approximately 100 years of urban development, now returning that Pb to the next generations of people unfortunate enough to be members of the urban poor. This new paradigm, verified by our recent research in deciphering the causes of a seasonal oscillation in children's blood Pb levels, points to a relatively simple and cost-effective way forward in reducing the Pb load for urban youth.

Effects of Pb on Humans

Compared to other chemicals of environmental concern, the uptake mechanisms and toxicological effects of Pb are relatively well understood. The primary pathway of Pb uptake in humans is via ingestion, where Pb is absorbed in the intestine and incorporated into the body (Manton et al. 2001). People's absorption potential for Pb is dependent mainly on age—the portion of ingested Pb that is taken up in the body is typically less than 5% for adults whereas it is as high as 50% for children due to their less-developed gastrointestinal pathway (Roberts et al. 2001). Once absorbed, Pb moves through a number of biological systems in the body. First, due to similar charges and ionic radii, Pb is utilized in biological processes much like calcium, including acting as a critical component in converting the electrical neural signal into a chemical signal and as a key mineral component of bone formation. If the Pb ends up in a neuron, it does not function as a neurotransmitter as calcium (Ca) does; instead it effectively creates permanent neural differentiation defects resulting in lowered IQ, learning disorders, and attention deficit hyperactivity disorder (Nevin 2000; Nigg et al. 2008). Because of their high ingestion efficiency and the rapid neural differentiation that occurs during early brain and nervous system development, children are especially vulnerable to permanent effects of Pb poisoning. When Pb is incorporated in bone material, the bone becomes a longer-term source of Pb to the biological system. Bone is regenerated on time scales of months to years, leaking additional Pb into the system. (Some evidence also suggests that elderly suffering from osteoporosis have elevated blood Pb levels from bone-loss related sources, which decreases cognitive function (Needleman 2004)). For this reason, children treated by medical interventions, such as blood chelation, may continue exhibiting toxic levels of Pb in their blood. Furthermore, as neither the placenta nor mammary glands are perfect barriers to Pb, pregnant and lactating mothers with elevated blood Pb levels may themselves pose a health risk to babies and fetuses.

The health standards for Pb levels in blood have been steadily revised downward over the years as medical research has shown toxicological effects of Pb in increasingly lower quantities. The U.S. Centers for Disease Control and Prevention (CDC) in 1991 chose 10 micrograms per deciliter (μg dL^{-1}) as an initial screening level for Pb in children's blood, although subsequent studies have been unable to determine a "safe" lower level of Pb, with levels below 10 μg dL^{-1} still causing some toxicological effects. The full spectrum of toxicological effects of Pb in the human system is still not known and deserves further study. But the persistent presence of Pb in children is a public health issue of first order.

The Lead Paint Paradigm

Pb has been used by humans for thousands of years and its toxicity has been known for centuries, but it was not until the Industrial Revolution that this issue became a widespread problem. Lead is a soft and malleable metal easily extracted from galena ore through simple heating. The Romans established a metal-based society early and intensively. Using the newly-conquered Iberian Peninsula, a target of the Romans in large part because of its rich metal ore deposits, the Romans developed the first large-scale quarrying and working operations for Pb, using the finished product in containers, water pipes, and as a sweetener in wines to counteract high tannin levels (Gilfillan 1965). The environmental legacy of Roman mining in Spain still plagues a number of regions with severe contamination problems. Lead production and use fell drastically following the Roman era, with some use in alloys, soldering, glazes, and containers, until the 20th century.

Pb has been added to paint for centuries—distinctive colors are achieved with the addition of metals to paints. But a boom in residential housing development in the early part of the 20th century resulted in a national-scale advertising blitz for "white lead paint" and the application of Pb-based paints in millions of new homes. The addition of Pb, in practice up to 15% by weight, enhanced durability and flexibility of paints. In the early part of the 20th century, most single- and multi-family dwellings had Pb-based paints in their walls, window sashes, and doorways. Even brick and stone houses employed Pb-based paints in windows and doors. Although Pb enhanced durability, paint has its functional limits, and the degradation around friction points (doorways, window sashes) combined with the exploratory nature and oral fixation of young children resulted in the first widespread tragedy of Pb poisoning. As children were being admitted to hospitals with symptoms of severe and chronic Pb poisoning, the link to Pb-based paints became apparent as the only significant source of Pb to these children. In the 1940s, pressure from the health profession and consumer advocate groups resulted in legislation prohibiting the addition of Pb to house paints. Although still allowed to this day in industrial applications like bridge paints, with potential implications for train trestle bridges in urban settings (Weiss et al. 2006), the banning of Pb in house paints in 1950 gave hope for a Pb-free future for children. This Pb-free future never came to fruition for two reasons: the explosion of automobile use after WWII, fueled by leaded-gasoline (much more on this later), and the inevitable degradation of Pb-based paint in and around homes.

The continued poisoning of children from Pb-based paints was a sadly predictable outcome. The fact that paint applied after 1950 was Pb-free didn't change the Pb content of old paint (e.g., Ter Haar and Aronow 1974). Anybody who has refinished an older home is aware of the problem—what do you do with the Pb paint on the walls, sills, and doorways? The popular way to refinish trimwork and windows is the most problematic. Sanding of Pb-based paints converts the paint from a glue-type solid with limited bioavailability into millions of fine particulates with relatively high Pb content and very high bioavailability, due to the high surface area/mass ratio of these particles. Among the acute cases of Pb poisoning seen in children today in the U.S., many of them are from refinishing/construction contact with Pb. This problem can bridge class and race—remodeling of older homes is often a luxury of the upper middle class, as they restore a historic home to its original luster.

To confront this problem, many health and environmental agencies at the national, state, and local levels have been waging a war of remediation and education about the hazards of Pb. Most of the remediation efforts have been focused on safely removing or covering Pb-based paints in homes—approximately 26% of all U.S. housing stock was built before 1950, and 24 million homes still contain Pb-based paint (HUD and BC 1999). These remediation efforts continue to this day, with almost $120 million allocated by the U.S. Department of Housing and Urban Development for Pb reduction projects in 2009 alone. The agencies involved have touted these efforts as a success, holding up the improvement in the number of children affected by Pb over the past 25 years. In a national health assessment survey in the late 1970s, 88% of our nation's children had blood Pb levels above that deemed safe by today's standards. In a follow-up survey in the 1990s, that number was down to 2.2%, with annual improvements seen in interim surveys up to today. When we ask people what they consider the key pathway for Pb to children, they invariably respond that kids get Pb-poisoned from eating paint chips.

When medical, scientific, and regulatory findings reach the collective psyche of society, a paradigm is formed. This paradigm, that Pb-based paints still constitute the biggest risk to children with respect to Pb, and that the remediation of Pb-based paint sources has in the past and will continue in the future to provide the chief benefits to children's health, is firmly entrenched. The seduction of this idea is easy to see—images of toddler's bite marks of painted window sills, X-rays revealing paint chips in a child's stomach, and a photo of a white-clothed team of remediation experts removing Pb-based paint from a building can be superimposed to create an image of a neat, clean and effective solution to this problem. This seduction has recently hit the courtroom, where several high-profile cases brought before juries have revolved around large paint producers, like Sherwin-Williams, who have been sued for producing Pb-based paints over 60 years ago. In the first major case, which appeared in Rhode Island Superior

Court, a six-person jury found Pb paint producers liable for creating a multi-billion dollar "public nuisance" by producing Pb-based paints applied to a quarter million houses in Rhode Island (Rabin 2006). This ruling was overturned just two years later by the Rhode Island Supreme Court. Corporate and industrial culpability should extend to producing products that knowingly endanger the health of people and the environment, and the idea of reparations to support remediation of this public health menace is certainly an appealing one, if the real scope of the "public nuisance" is truly understood.

But what if the paradigm is wrong? What if poor urban youth are no longer being poisoned by chipping paint, but rather the soil around them? What if the vision of a white-clad team of specialists sweeping through a housing project, removing Pb from the walls and leaving in its wake sparkling new Pb-free paint, needs to be replaced by an image much more messy and comprehensive to solve this problem of social injustice? How will we know when to abandon the Pb-based paint paradigm for another one? We believe that the time is now, and is bolstered by a series of findings that are inconsistent with the Pb-paint only paradigm for the poisoning of our children by Pb. In particular, the inability of remediation of paint alone to reduce the blood Pb level of urban youth is one key clue that we are missing a critical additional source of Pb to our children.

A New Paradigm: Soils as the Vehicle for Continued Pb Poisoning of Urban Youth

We have hit the wall in terms of improving the Pb-poisoning outlook for some children (e.g., Fig. 1.1), particularly for those of color living at or below the poverty level in older cities. Even after decades of active intervention, these urban youth have Pb-poisoning rates that are up to 10 times the national average. In 1994, a summary statement from a national health survey stated that the exposure to Pb at levels that may adversely affect the health of children remains a problem especially for those who are minority, urban, and from low-income families. Strategies to identify the most vulnerable risk groups are necessary to further reduce Pb exposure in the United States. These socio-economic risk factors include poor nutrition with the potential for pica behavior (a subconscious desire to ingest soil and dust to overcome nutritional deficits), and inadequate pediatric health care. Additionally, and of critical importance for improving the health outcome of urban youth, these risk factors also include poor home maintenance (which correlates with high rental percentages), high dust and dirt exposure (which correlates with high percentages living in urban housing), and relatively low awareness of the links between health and behavior. In particular, the continued poisoning of urban youth from the very dirt and soil upon which they live is the key to a new emerging paradigm—namely, that the continued source of Pb to children is Pb-enriched soils that are prevalent in cities, especially older ones. The source of Pb to the soils includes degraded Pb-based paints, but also Pb deposited from tailpipes, the result of 60 years of combustion of leaded gasoline. In fact, much of the blame for chronic Pb poisoning and for the improvement in the national average of blood Pb may be the banning of Pb as an additive in gasoline in 1980.

The production and use of Pb for gasoline additives was spurred by the need to control the explosion of gasoline in cylinders of internal combustion engines. The formulation of tetraethyl Pb as a fuel additive was perfected in the 1920s, resulting in the adoption of a global fuel standard that contained about 2% Pb oxide by weight. An early warning sign went up when scores of workers were severely poisoned in the 1920s by Pb toxicity in plants producing tetraethyl Pb additives, although a multi-pronged industrial cover-up limited public awareness of this situation. But the dawn of the automobile age shelved concerns of the environmental impacts of tetraethyl Pb as affordable transportation dramatically altered the American landscape of the 20th century. Lead use for this application followed the trend in automobile use in America, with 250,000 tons of Pb used in gasoline and emitted by tailpipes every year by 1970.

Roadway sources of Pb

Overall, about 5 million metric tons of Pb were deposited in the environment as a result of the combustion of leaded gasoline (Fig. 1.1). Almost all of that Pb was originally deposited very close to roadways, with aerosolized combustion products containing Pb initially deposited within about 50 m of a roadway if no obstructions were present. The fate of deposited Pb then depended on the conditions of the depositional area. Although intersections of busy streets may have received over 1 metric ton of Pb per year, their impervious surfaces led to continual runoff of Pb-enriched particulates down storm drains (and from there into treatment plants or directly into rivers). If the particulate Pb was deposited instead on a grassy fringe, like a front yard or park, the Pb was effectively retained. In such a setting, the insolubility of Pb leads to surface peaks in Pb concentration of soils; in relatively undisturbed soils, this surface Pb enrichment may be the product of decades of Pb deposition from gasoline and reach levels above 1000 parts per million (ppm). Thus, surface soils became the repositories of Pb deposited over decades—in the case of older roadways, the proximal soils retained almost all of the Pb deposited on them over a period of about 60 years.

The roadway Pb is generally bioavailable, being present in carbonate and oxyhydroxide soil fractions, while the Pb in natural soils is relatively inert. Therefore, dust originating from urban soils contaminated by anthropogenic Pb is more toxic than naturally occurring Pb dust (Chlopecka et al. 1996; Lee et al. 1997). Lead from the combustion of leaded gasoline is preferentially enriched in the more readily windblown fine size fraction of soils, and so Pb in dusts derived from urban soils is likely to be more potent and concentrated that in the bulk soils (Young et al. 2002).

Diffuse soil Pb and children's health

The original sources of Pb to the environment were directly tied to the spatial characteristics of the product itself, with Pb-based paints linked to pre-1950 structures, gasoline-related Pb linked to roadways and traffic volume, and Pb emitted from smelters linked to the smelter location, stack height, and wind direction. As detailed above, Pb does not originally deposit far from its source, and its geochemical characteristics promote rapid sequestration onto surface soil particulates (usually via surface complexation of Pb and Pb oxides with soil organic matter). But an analysis of many urban areas reveals that these point sources have, to some extent at least, been redistributed to produce regions of Pb enrichment. Several factors can lead to redistribution of Pb-enriched particles and soil, but the recurrence of a general urban enrichment of soil Pb has been documented in many regions, and is termed diffuse soil Pb.

One of the characteristics of Pb distribution in surface soils of several older cities is a distinct decreasing trend from city center to suburban surroundings, a legacy both of Pb deposition, redistribution and smearing of original point sources, and less Pb deposition in newer suburban neighborhoods due to recent Pb controls (e.g., Mielke et al. 1984/85; Filippelli et al. 2005; Johnson and Bretsch 2002; Roux and Marra 2007). The urban roadway example shows both the impact of the point source of Pb deposition from leaded gasoline as well as the diffuse soil Pb that blankets urban regions. In other words, even at distances away from the roadway beyond where direct Pb deposition occurs (and far away from structures using Pb-based paint), the background level for Pb is significantly higher in urban areas (~500 ppm) than in suburban areas (~ 60 ppm) (Laidlaw and Filippelli 2008). This urban-suburban gradient is one overriding factor affecting the amount of Pb loading to individuals, a factor that we will next assess on a larger scale and with respect to human health.

The actual distribution of people with blood Pb levels exceeding action limits is getting more difficult to obtain due to privacy issues, but in the past, blood Pb values could be collected from health department records down to the level of a street address, providing an outstanding way to examine the environmental factors in human health. The distribution of blood Pb levels exceeding action limits in Indianapolis from 1992-1994 is informative. Many of the higher blood Pb values were concentrated in urban areas, particular the downtown. In contrast, very few incidences of blood Pb poisoning were found in the newer suburban areas to the west, north, south, and northeast sides of the city. Because these were individual blood Pb data, population density plays some role in the distribution; for example, rural farmlands on the city outskirts have few incidences. But based on the 1999-2000 U.S. Census, the population density per census tract in the newer suburban areas with few blood Pb poisoning incidences is comparable to the urban and near-urban areas with a high incidence.

To explore the concept of diffuse soil Pb and its potential role in affecting children's health in Indianapolis, we performed a coupled soil survey and comparative epidemiological analysis. Our soil sampling criteria included being greater than 50 m from roadways and from structures (which might have contributed Pb-based paint), but was augmented by aerial photographic records over Indianapolis from several time slices from 1940 to 1970. The purpose of these aerial photographs was to rule out the potential for inadvertently sampling soils from disturbed, excavated, or filled areas that might have surface Pb contents characteristic of artificial materials rather than natural soil. As one can imagine in a rapidly developing urban area, this criteria narrowed down acceptable sites to only about 100 distinct sites, many of which were in parks, cemeteries, and school grounds. Analyses were also carried out to determine whether soil source material showed any inherent Pb variation. No trend was found between Pb content and soil composition across Indianapolis, and thus we concluded that soil mineralogy is a minor control on the Pb distributions presented here.

In contrast to detailed roadway and house-side soil sampling, which revealed Pb concentrations above 1000 ppm (Fig. 1.2), the highest soil Pb concentrations in our study were below 500 ppm (Fig. 1.2). The lowest Pb concentrations averaged about 50 ppm, which is a typical value for soils in this region. As expected, the highest soil Pb concentrations were focused in a bulls-eye pattern directly over the old urban areas of Indianapolis, where the diffuse soil Pb content averaged over 200 ppm. Beyond this central hot spot, Pb concentrations decreased systematically toward the suburban outskirts of the city, ultimately falling to background values in the rural fringes of the city. The central peak is consistent with the long history of Pb use in the downtown, but the generally high values even away from point sources support the argument of a redistribution of Pb over time. This is a common feature of urban Pb distribution, and is likely related to the wind-driven redistribution of fine Pb-enriched particulates in a statistically consistent pattern over decades. We do not see a "plume" of deposition that can be ascribed to the northwestward prevailing winds in Indianapolis, likely because, unlike releases of particulates at higher elevations (i.e., smokestacks), wind direction has little influence on dust depositional patterns in the turbulent near-surface environment of a cityscape.

Although many factors influence the relationship between geology and human health in the story of Pb, it is clear from the lack of closure on this issue that we do not yet understand all of the confounding factors. Furthermore, this generalized approach presented above provides a reference point for further work, but does not integrate well health data and geologic data, nor does it present recommendations that Earth scientists can make to health specialists in further reducing this public health hazard beyond the incredibly costly and disruptive solution of removing all of the contaminated surface soil in urban areas and replacing it with clean fill. Several bridging efforts are now being pursued to help further medical geology in the context of eliminating childhood Pb poisoning. Beyond simply documenting Pb distribution and its public health implications, current research is also examining more closely Pb as a toxicological agent with predictable behavior. For example, isotopic techniques have been utilized to closely examine the entry mechanisms of Pb into the body and the cycling of Pb within the body, with a goal of pinpointing Pb toxicity in individuals and thus more closely coupling prevention and treatment. Another new tool of promise in accurately assessing Pb poisoning is predictive modeling of children's blood Pb levels using weather data.

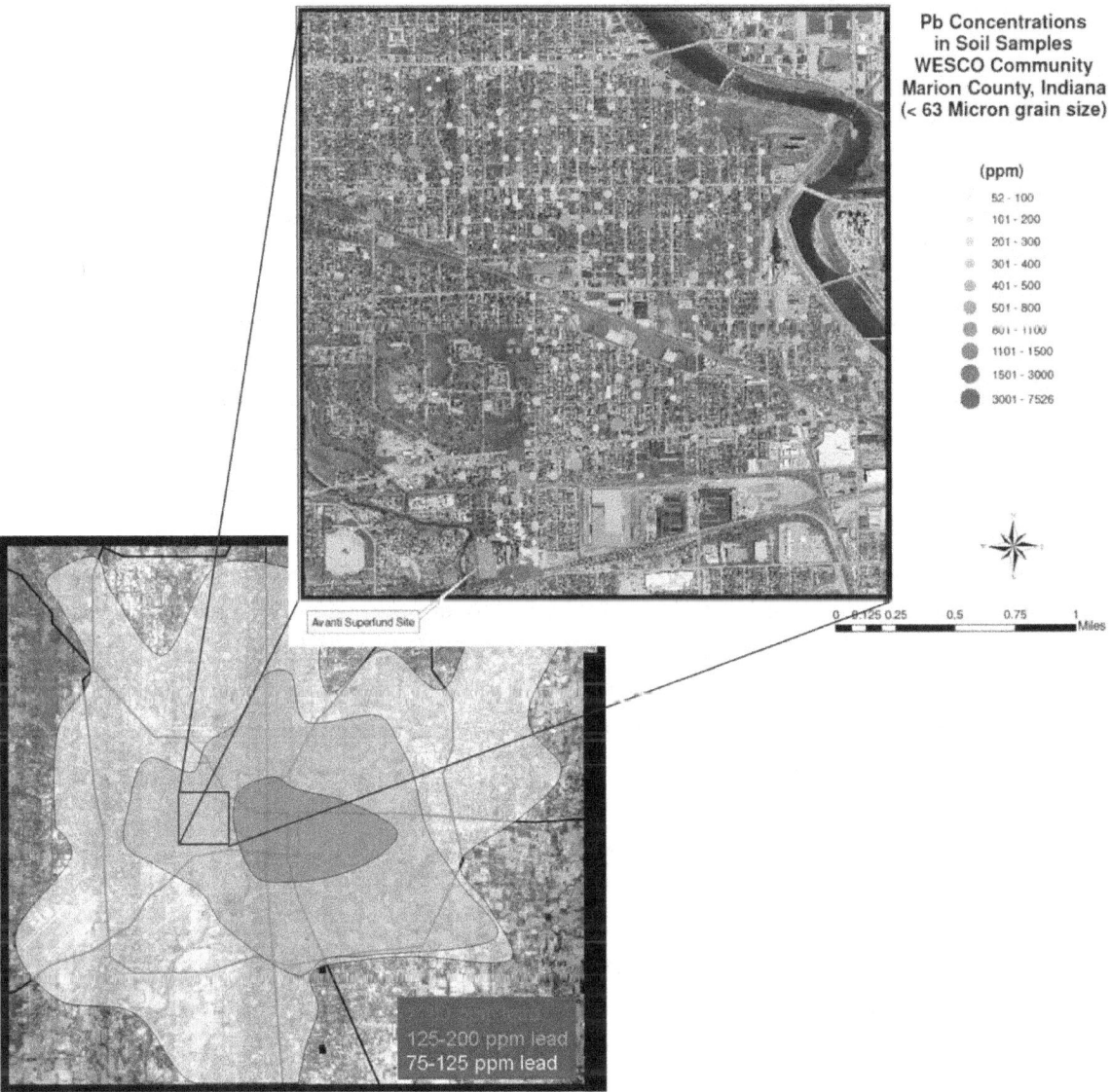

Diffuse soil Pb >40 m from roadways and structures

Fig. 1.2. Diffuse soil Pb (measured more than 40 m from roadways and structures) in Indianapolis, Indiana (lower left) and detailed blow-up of higher resolution yard-scale sampling in a neighborhood with high incidence of elevated blood Pb levels (upper right). The concentration of diffuse soil Pb in surface soils of Indianapolis (in colored regions) displays a characteristic pattern of urban enrichment trending toward background values in suburban and agricultural regions. (Color version is available in the EBook or from the author at StevenE@uvu.edu.)

A New Paradigm Confirmed? Seasonal Factors and a Blood Pb Predictive Model for Health Care Research

Several studies have identified a seasonal trend in blood Pb levels (e.g., Billick et al. 1979; Rabinowitz and Needleman 1982; Rothenberg et al. 1996; Haley and Talbot 2004; Laidlaw et al. 2005), with average monthly blood Pb levels of children from urban areas increasing significantly in summer months. Summer increases of children's blood Pb levels were so prominent over many years in Syracuse, New York, that a group of researchers led by David Johnson at the State University of New York, College of Environmental Science and Forestry, concluded that the phenomenon is probably caused by the interaction between climate and soils, leading to enhanced dust Pb loading to children (Johnson et al. 1996). An intriguing alternative hypothesis for blood Pb seasonality is internal, whereby bone material is increasingly recycled during summer months, releasing stored Pb to the blood stream.

To better constrain this climate/soil/human health link, we have been investigating in detail variations in children's blood Pb levels as a function of climate and soil factors in several urban areas. The ultimate goal of this effort is to develop a predictive model whereby a medical researcher can make an accurate diagnosis of Pb poisoning based on seasonal and weather-related factors as well as blood Pb level data. We used a number of independent climatological variables, including average monthly soil moisture, PM10 (particulate matter less than 10 microns in size), wind speed, and temperature obtained from state and federal government data sources. We also used blood Pb databases obtained from local and state governmental sources. The average blood Pb concentration was computed using the child blood Pb measurements for each month. The outcome variable, children's average monthly city blood Pb concentration, was regressed against the average monthly independent variables soil moisture, PM10, wind speed, temperature, interaction variables, and monthly dummy variables using backward elimination procedures. A wind rose of Indianapolis shows dominant wind directions from the west to northwest, but in our initial analysis wind direction had no predictive application for blood Pb values.

This model indicates that the variables or interaction variables including soil moisture, wind speed, PM10, temperature, and the monthly dummy variables for March, April, June, July, August, and September, explain 87% of the variation in the response variable, monthly average child blood Pb level concentrations (Fig. 1.3). Based on this multiple regression model, and recently published results from several other American cities (Laidlaw and Filippelli 2008), we believe that the seasonality in children's blood Pb levels is controlled by exposure to Pb dust originating from contaminated soils and suspended in the air when several weather related environmental conditions are present: temperature is high, soil moisture is low, and atmospheric PM10 is elevated. Under these combined weather conditions, Pb-enriched PM10 dust disperses in the urban environment and is manifest by elevated Pb dust loading. In this case, exposure is via increased dust loads in homes and on contact surfaces, with ingestion being the assumed uptake mechanism (although direct pulmonary uptake cannot be ruled out at this point, and is a seriously under-studied process). Thus, the critical factor is not so much the particular season, as we have demonstrated in New Orleans (Laidlaw et al. 2005), but rather drying out of surface soils and deflation of these soils in wind events that causes high Pb exposure. Although further work using detailed tracking of Pb, possibly involving Pb isotopic studies, may help to elucidate the connection between seasonality and blood Pb values, the ability of geochemical and meteorological factors to predict blood Pb supports the supposition that external loading and exposure drive much of the blood Pb concentrations. The fact that the model predicts 87% percent of the variance in blood Pb indicates that, on the population scale at least, socioeconomic status is not the dominant factor in predicting blood Pb levels of children. Because resuspension of Pb from contaminated soil appears to be driving seasonal child blood Pb fluctuations, Pb contaminated soil in and of itself may be the primary driving mechanism of child blood Pb poisoning in the urban environment.

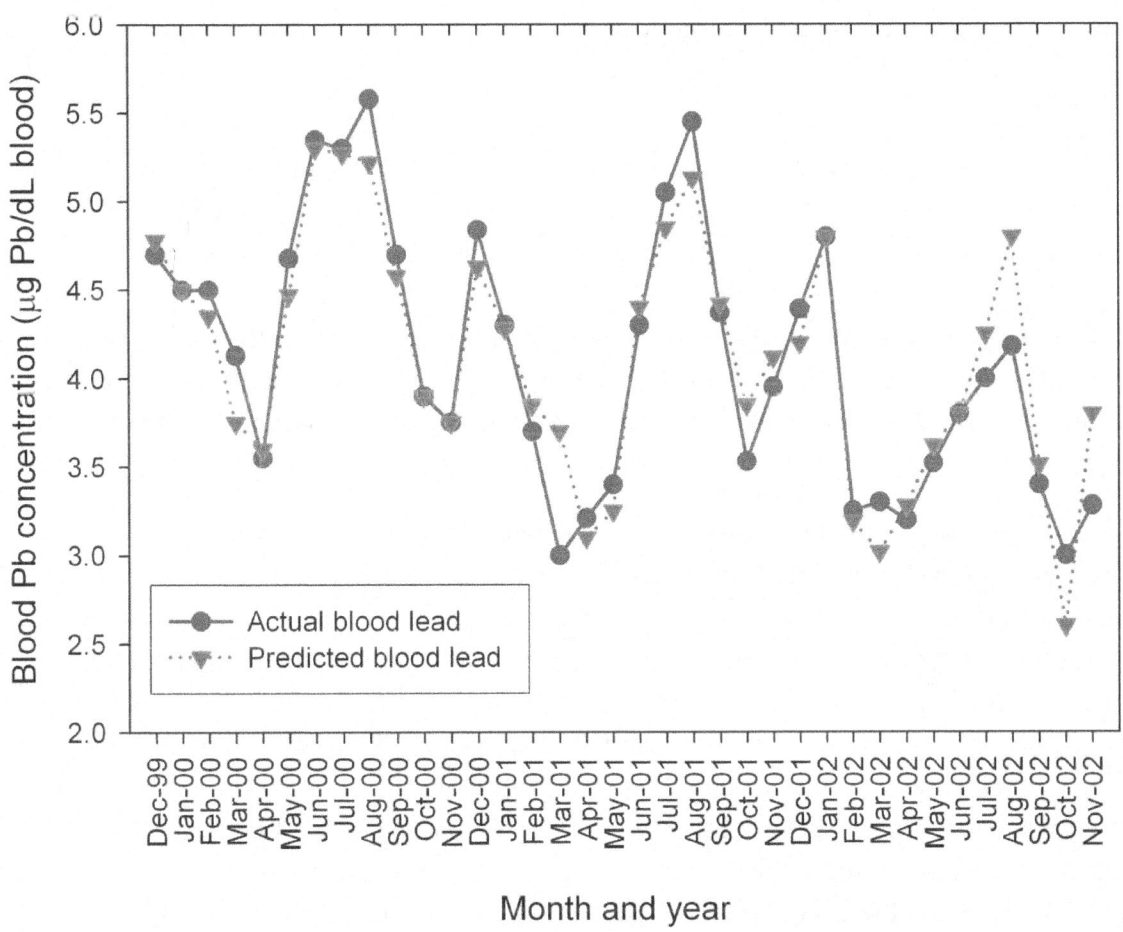

Fig. 1.3. Best-fit model results to predict blood Pb levels in children from Indianapolis based on meteorological factors alone (after Laidlaw et al. (2005)). The time period of the analyses consisted of 36 months between December 1999 and November 2002. The dependent variables for the models included monthly child blood Pb data from (1) a variety of subregions in Indianapolis, (2) a variety of blood Pb levels, and (3) a variety of ages (i.e., 0-1.0, 1.01 to 2.0, 2.01 to 3.0, 3.01 to 4, 4.01 to 5, and 5.01 to 7.0 years). The blood Pb database totals during this time interval included a monthly child blood Pb dataset of 15,969 children. This model indicates that the variables or interaction variables including soil moisture, wind speed, PM10, temperature, and the monthly dummy variables for March, April, June, July, August, and September, explain 87% of the variation in the response variable, monthly average child BLL concentration ($r^2 = 0.87$, p = 0.0004, n = 15,969). (Color version is available in the EBook or from the author at StevenE@uvu.edu.)

In addition to the development of hypotheses related to the incorporation of Pb into children's systems, a promising result of these modeling analyses is the ability to predict toxicity in a given population. In other words, through easily collected atmospheric and soil data, a health researcher can determine the expected variation in blood Pb levels of the general population and, if performed in more detail using the subset of children's age, the expected variation in a given young patient. This is particularly important when attempting to treat blood Pb poisoning using discrete venous sampling events—a safe level measured in the spring under conditions of high soil moisture could become a poisonous level in the same patient just several months later when atmospheric conditions increase ambient Pb loading.

Final Thoughts

In summary, a new paradigm of urban Pb loading is emerging, one that helps to explain continued chronic Pb poisoning and seasonal patterns in blood Pb levels of children. Unlike discrete point sources like Pb paint and industrial contact, which are still responsible for most cases of acute Pb poisoning, diffuse soil Pb is the main avenue for urban Pb loading of children. The diffuse soil Pb comes from several sources, including leaded gasoline and degraded Pb-based paints, but in a sense the source no longer matters—because of the ability of surface soils to retain Pb, these soils themselves have become the new risk factor for children's health in Pb-loaded cities.

Widespread contamination of urban soils creates a different challenge for mitigation of Pb risks for children, one based on removing surface soils from human contact. Most mitigation efforts for heavily-contaminated soils have involved soil removal and replacement, a disruptive and expensive option for controlling Pb sources in urban areas. Recently, another approach was tested (Mielke et al. 2006) in New Orleans that is much cheaper and appears to be as effective as soil removal. The approach was simply to cover the contaminated yard soils with about 15 cm of Pb-free soil, which in the case of New Orleans came from the nearby Mississippi levee. At a fraction of the soil removal cost, this clean soil is simply graded over the old soil layer, hydroseeded, and left to grow a lawn. This approach caps the Pb-contaminated soils, removing them from contact by children. The result of initial work is a substantial reduction in the blood Pb levels of children living in the affected homes. In a just-published paper, Zahran et al. (2010) report on how nature did this same experiment, producing substantially lower blood Pb levels for New Orleans children after Hurricane Katrina, due to the capping of much of the Pb-contaminated soil with flood-related sediments. Mielke et al. (2006) observed that, over the course of several months after treatment, soil Pb levels in the treated sites began increasing. This increase was due to the resuspension and deposition of soil from adjacent, untreated yards and neighborhoods that still had high soil Pb concentrations. This finding agrees with results from an urban gardening study in Boston (Clark et al. 2008), which revealed that raised beds experienced substantial increases in soil Pb values over as little as four years after bed construction, indicating the need to control soil Pb resuspension at the neighborhood scale. Collectively, these findings not only confirm the new paradigm of diffuse soil Pb as a culprit in urban areas, but also indicate that a comprehensive treatment approach is required to provide a long-term benefit.

The strategy of mitigating Pb risks in entire cities is of course different than doing such at a single contamination site or mine. In many ways, it seems more daunting because of the scale, but it is quite approachable given a newer understanding of urban Pb exposure sources. As indicated above, surface capping of Pb-contaminated soils is effective, especially if done on a neighborhood rather than individual property scale. Mielke et al. (2006) calculated that the entire Pb-affected area of New Orleans could be remediated by capping for a total cost of less than $300M, compared to the $76M annual cost to New Orleans of Pb poisoning. Simply put, the national cost of doing nothing about soil Pb is staggering—Gould (2009) calculated that Pb poisoning costs the nation $30-146M in special education, $267M in attention deficit-hyperactivity disorder, and $1.7B in the direct cost of crime and recidivism. The cost:benefit ratio of Pb reduction is favorable from a financial standpoint, but the necessity of reducing Pb burdens to urban youth goes far beyond dollars and cents, and continues to be a critical component of enhancing public health and social justice. Until we face the new paradigm that diffuse soil Pb is causing continued chronic Pb poisoning of urban youth, we will not be able relieve the environmental insults to these children, nor redress the social injustice that is the product of a century of Pb loading to US cities.

References

Billick, I. H., A. S. Curran, and D. R. Shier, 1979. Analysis of pediatric blood lead levels in New York City for 1970-1976. Environmental Health Perspectives, v. 31, pp. 183-190.

Chlopecka, A., J. R. Bacon, M. J. Wilson, and J. Kay, 1996. Forms of cadmium, lead, and zinc in contaminated soils from southwest Poland. Journal of Environmental Quality, v. 25, pp. 69-79.

Clark, H., D. Haustaden, and D. Brabander, 2008. Urban gardens: Lead exposure, recontamination mechanisms, and implications for remediation design. Environmental Research, v. 107, pp. 312-329.

Filippelli, G. M., M. A. S. Laidlaw, J. C. Latimer, and R. Raftis, 2005. Urban lead poisoning and medical geology: an unfinished story. GSA Today, v. 15, pp. 4-11.

Gilfillan S. C. 1965. Lead poisoning and the fall of Rome. Journal of Occupational Medicine, v. 7, pp. 53-60.

Gould, E., 2009. Childhood lead poisoning: Conservative estimates of the social and economic benefits of lead hazard control. Environmental Health Perspectives, v. 117, pp. 1162-1167.

Haley, V. B. and T. O. Talbot, 2004. Seasonality and trend in blood lead levels of New York State children. BMC Pediatrics, v. 4, p. 8.

HUD and BC (U.S. Department of Housing and Urban Development and the Bureau of the Census), 1999. American Housing Survey for the United States. Current Housing Reports, Series H150/99, U.S. Government Printing Office, Washington, D.C.

Johnson, D., K. McDade, and D. Griffith, 1996. Seasonal variation in pediatric blood levels in Syracuse, NY, USA. Environmental Geochemistry and Health, v. 18, pp. 81-88.

Johnson, D. L. and J. K. Bretsch, 2002. Soil lead and children's blood lead levels in Syracuse, NY, USA. Environmental Geochemistry and Health, v. 24, pp. 375-385.

Laidlaw, M. A. S. and G. M. Filippelli, 2008. Resuspension of urban soils as a persistent source of lead poisoning in children: A review and new directions. Applied Geochemistry, v. 23, pp. 2021-2039.

Laidlaw, M. A. S., H. W. Mielke, G. M. Filippelli, D. L. Johnson, and C. R. Gonzales, 2005. Seasonality and children's blood lead levels: developing a predictive model using climatic variables and blood lead data from Indianapolis, Indiana, Syracuse, New York, and New Orleans, Louisiana (USA). Environmental Health Perspectives, v. 113, pp. 793-800.

Lee, P.-K., J. C. Touray, P. Baillif, and J. P. Ildefonse, 1997. Heavy metal contamination of settling particles in a retention pond along the A-71 motorway in Sologne, France. Science of the Total Environment, v. 201, pp. 1-15.

Manton, W. I., C. R. Angle, K. L. Stanek, Y. R. Reese, and A. J. Kuehnemann, 2001. Acquisition and retention of lead by young children. Environmental Research, v. 82, pp. 60-80.

Mielke, H. W., 1994. Lead in New Orleans soils: New images of an urban environment. Environmental Geochemistry and Health, v. 16, pp. 123-128.

Mielke, H. W. 1999. Lead in the inner cities. American Scientist, v. 87, p. 62.

Mielke, H. W., S. Barroughs, R. Wade, T. Yarrow, and P. W. Mielke, Jr., 1984/85. Urban lead in Minnesota: Soil transect results of four cities. Journal of the Minnesota Academy of Sciences, v. 50, pp. 19-24.

Mielke, H. W., E. T. Powell, C. R. Gonzales, R. T. Ottesen, and M. Langedal, 2006. New Orleans soil lead (Pb) cleanup using Mississippi River alluvium: Need, feasibility, and cost. Environmental Science and Technology, v. 40, pp. 2684-2789.

NRC (National Research Council), 1980. Lead in the human environment. National Research Council, Committee on Lead in the Human Environment, Washington, D.C.

Needleman, H., 2004. Lead poisoning. Annual Reviews of Medicine, v. 55, pp. 209-222.

Nevin, R., 2000. How lead exposure relates to temporal changes in IQ, violent crime, and unwed pregnancy. Environmental Research, Section A, v. 83, pp. 1-22.

NHANES (National Health and Nutrition Examination Survey), 2003-2006. National Health and Nutrition Examination Survey. U.S. Department of Health and Human Services, U.S. Centers for Disease

Control and Prevention, National Center for Health Statistics, Hyattsville, MD. Available online at http://www.cdc.gov/nchs/nhanes.htm.

Nigg, J. T., G. M. Knottnerus, M. M. Martel, M. Nikolas, K. Cavanagh, W. Karmaus, and M. D. Rappley, 2008. Low blood lead levels associated with clinically diagnosed attention deficit/hyperactivity disorder and mediated by weak cognitive control. Biological Psychiatry, v. 63, pp. 325-331.

Rabin, R., 2006. The Rhode Island lead paint lawsuit: Where do we go from here? New Solutions, v. 16, pp. 353-363.

Rabinowitz, M. B. and H. L. Needleman, 1982. Temporal trends in the lead concentrations of umbilical cord blood. Science, v. 216, pp. 1429-1431.

Roberts, J. R., J. R. Reigert, M. Eberling, and T. C. Hulsey, 2001. Time required for blood lead levels to decline in nonchelated children. Journal of Toxicology and Clinical Toxicology, v. 39, pp. 153-160.

Rothenberg, S. J., F. A. Williams, Jr., S. Delrahim, F. Khan, M. Kraft, M. Lu, M. Manolo, M. Sanchez, and D. J. Wooten, 1996. Blood levels in children in South Central Los Angeles. Archives of Environmental Health, v. 51, pp. 383-388.

Roux, K. E. and P. P. Marra, 2007. The presence and impact of environmental lead in passerine birds along an urban to rural land use gradient. Archives of Environmental Contamination and Toxicology, v. 53, pp. 261-275.

Ter Haar, G. and R. Aronow, 1974. New information on lead in dirt and dust as related to the childhood lead problem. Environmental Health Perspectives, v. 7, pp. 83-89.

Weiss, A. L., J. Caravanos, M. J. Blaise, and R. J. Jaeger, 2006. Distribution of lead in urban roadway grit and its association with elevated steel structures. Chemosphere, v. 65, pp. 1762-1771.

Young, T. M., D. A. Heeraman, G. Sirin, and L. L. Ashbaugh, 2002. Resuspension of soil as a source of airborne lead near industrial facilities and highways. Environmental Science and Technology, v. 36, pp. 2484-2490.

Zahran, S., H. W. Mielke, C. R. Gonzales, E. T. Powell, and S. Weiler, 2010. New Orleans before and after Hurricanes Katrina/Rita: A quasi-experiment of the association between soil lead and children's blood lead. Environmental Science and Technology, v. 44, pp. 4433-4440.

Chapter 2: Fishing on the Detroit River: Historical Pollution, Environmental Racism, and the Scientific State

Victoria Kalkirtz, Michelle Martinez and Alexandria Teague

Summary

The Detroit River is a vital resource to the City of Detroit and the region. From the earliest inhabitants, the river has been used as a source of food and community. Yet as the region went through a rapid period of heavy industrialization, so did the Detroit River. As pollution levels increased the water's toxicity, so did the level of contamination of many animals dependent on the river, especially fish. It remains in question whether it is "safe to eat the fish," particularly in urban areas. Although there have been some attempts at creating regional fish consumption advisories, testing procedures and dissemination of information are left up to a patchwork of city, county, state, provincial, and national agencies. Furthermore, economic and institutional constraints prevent local and federal governments from making the advisories readily available, which could be done by distributing them through local bait shops, retail outlets, and other local community centers. The combination of lack of access to fish advisories, history of water contamination and cultural/dietary needs escalates the environmental justice issue of consuming fish from the Detroit River. In an era of budget deficits, creative solutions must include collaborative partnerships between state and non-governmental actors and between scientists and community activists that speak beyond institutional and epistemological divisions. Environmental scientists can foster environmental justice by coordinating the collection and dissemination of data on the risks of fish consumption and by instituting community-based outreach and participatory research.

Introduction

The Detroit River is a vital resource to the City of Detroit and the region. From the earliest inhabitants, the river has been used as a source of food and community. Yet as the region went through a rapid period of heavy industrialization, so did the Detroit River. Over 100 years, the river accelerated into an entity of production, extraction, distribution, and waste disposal. As pollution levels increased the water's toxicity, the contamination level of many animals dependent on the river, especially fish, increased as well. Simultaneously, while the Detroit River was subject to change, so was the social structure of the City of Detroit itself, where over 100 years, it amalgamated into one of de facto segregation and economic deprivation. And still the river remains a vital source of food and community for Detroiters, a city composed of approximately 90% people of color. Despite the centuries of heavy industrialization along the shores of Canada and the United States, many anglers of color, and/or anglers of little means continue to use the resource for fresh walleye, yellow perch and catfish, flocking to the Detroit River to extract high sources of protein and Omega-3s. Others go to the river for the quiet that they enjoy and the sense of community fishing offers. This complex history puts many anglers and those who consume fish from the river in a quandary of environmental injustice.

"Is it safe to eat the fish?" remains an open question, especially in the urban areas. The State of Michigan currently has a fish consumption advisory on six species in the Detroit River, limiting their consumption due to chemicals such as poly-chlorinated organic compounds (PCBs), mercury, and dioxin, all known to have harmful effects on human health ranging from carcinogenic properties to properties that affect human development. On the Canadian side of the river, you will find their warning against consuming nine different species from the Detroit River, depending on where you fish. The Canadian advisories are based on many of the same chemicals as the Michigan advisories, but also include mirex, photomirex, DDT, and toxophene from chemical fertilizers. The health benefits of fish consumption are great and widely understood as demonstrated through dozens of studies on the matter. Omega-3 fatty acids provide many positive health effects including cardiovascular and cognitive benefits. These positive health effects are increased for pregnant women and their children (Sidhu 2003).

This conundrum created a need to balance the benefits and risks of eating fish, caught both commercially and through sport fishing, a dilemma dealt with through state agencies. This is especially true of those who rely on fish for subsistence or as a significant supplement to their diets. Many urban residents utilize polluted waters as their fishing areas, thus increasing their risk of ingesting contaminated fish. Fish advisories are considered just that, advice for the fish consumer to make informed decisions while weighing the risks and benefits of eating potentially contaminated fish. Yet, quite often the vulnerable populations not only do not receive this information, but also are unable to follow the advice due to dire economic and food security issues. Further, people fish because it is of high value from social, psychological, dietary, cultural, and intergenerational standpoints. Economic and institutional constraints prevent local and federal governments from making the advisories readily available, which could be done by distributing them through local bait shops, retail outlets, and other local community centers. Additionally, these and other regulatory constraints prevent the reduction and elimination of the sources of pollution, which generates the necessity to offer advice rather than take action. The combination of lack of access to fish advisories, history of water contamination, and cultural/dietary needs escalates the environmental justice issue of consuming fish from the Detroit River.

Detroit's Industrial History: A Brief Introduction

The historical uses of the Detroit River as a natural and industrial resource assist in understanding the current state of the river, environmental justice and fish consumption. The tradition of fishing on the Detroit River is one that began about 6000 B.C. Early French explorers describe at least seven Native American tribes that inhabited the area: the Huron (Wyandot), Ottawa, Chippewa (Ojibwa), Fox, Sac (or Saulk), Miami and Potawatomi (Anishnabe). The picture painted is one of swampy marshlands, commerce and trade throughout the "meicigama" or "great water" in Ojibwa. In 1815, the number of Native Americans in Michigan was about 40,000; in 1825 nearly 30,000; in 1880 there were only 10,141 (Woodford 2001).

Simultaneously, the City of Detroit, the capital of Michigan from 1824 to 1850, sustained an influx of people that numbered 1,650 in 1810, rapidly grew to 45,600 in merely 50 years, and reached 285,704 at the turn of the century. This shift in population consequently changed the definition and use of the Detroit River as a resource. It accelerated the river's focus to a mechanism of commercial and industrial growth, ultimately resulting in the conflict around the adequacy of its use as a food resource. The steady inflow of people throughout the century was the result of a deliberate accession of land and recruitment of business to make Detroit a vital commercial and cultural center. From 1813 to 1831 Lewis Cass, territorial governor of Michigan, recruited settlers in southeast Michigan with campaigns advertising productive farmland, which had a reputation of being swampy and inadequate for farming, in hopes of boosting the economy. The first steamboat on the upper Great Lakes, the opening of the Erie and Soo Canals, and the Michigan Central Railroad all allowed traffic from New York to flow to all major Great Lakes cities, solidifying Detroit as a major center for commerce and culture by the mid-nineteenth century (Glazer 1965; Kerr et al 2003).

The availability and abundance of raw materials facilitated the growth of manufacturing. While diverse sources of income facilitated the creation of a vibrant economy through exporting raw and processed goods, importing new laborers sustained such growth (Glazer 1965). Detroit became one of the largest producers of railroad cars, ships, and heating and cooking stoves (Dunbar 1965; Hyde 1980). Detroit also became known for chemical production, tobacco processing, and pharmaceutical industries along with over a dozen other small industries (Hyde 1980). From Chicago to New York City, Detroit became a major source of goods in merely 100 years. Detroit machine shops and foundries produced metalworking skills that would lay the foundation for the next era of industrial growth, the automotive industry (Hyde 1980).

The story of the Model-T is indelibly marked in history books as it revolutionized labor and production for the world. In 1903, Henry Ford had established the Ford Motor Company, selling an unprecedented 15 million Model Ts by 1927. The wages rose from $2.16 a day in 1913 to $5.13 a day in 1919, and prior to WWI, the automotive industry employed 120,000 persons. From a population of 285,704 in 1900, Detroit attracted hundreds of thousands of people, in thirty years adding to a total of 1,568,662 (Glazer 1965). This era was also responsible for the strengthening of the UAW, the renaissance of Motown music, and the cultural pride of the city. Detroit became the Motor City and one of the largest industrial manufacturers in the world.

Today, the Detroit River forms the foundation for the cities of Detroit, Michigan and Windsor, Ontario, among eight more municipalities. Named Le Detroit, the French word for "strait", the Detroit River provides a throughway between the Great Lakes of Huron and Erie. The Detroit River is 32 miles long, an average of 35 feet deep, and scattered with 15 islands. The largest of these is Belle Isle, a park since 1879, and Grosse Ile, a township of 10,000 residents (Woodford 2001). Detroit, like many cities that are defined by neighboring water bodies, exists not only because of it, but also relies on it for transportation, shipping, water, sewage treatment and industrial purposes. As Detroit became more of an industrial city, the river was increasingly used as a means for transportation of goods and disposal of waste (Kerr et al. 2003). The externalities of commerce and historical contamination laden on the Detroit River still have far reaching implications for human and ecological health.

Changing Demographics and City Development: Race, Class and Environmental Inequity

As the use of the Detroit River shifted to provide water for the many industrial processes, a simultaneous re-structuring of labor, race and class also defined how those various social sectors would interact with the environment and the Detroit River. Throughout the period of industrial growth, racial demographics, class structure and political struggles began to form that would define the path of Detroit's history for the entrance into the 21st century. Historian Olivier Zunz (2000) writes that, during industrialization, Detroit changed from self-sufficient, multi-class ethnic neighborhoods of the 19th century into socio-ethnic specific neighborhoods divided by urban and suburban zoning. In this era, it was not only the degree to which the city grew, but also the nature of the population concentration within the city that made way for a new formation around class and race. It was also in this era of rapid industrialization that the arrival of African Americans in large numbers resulted in their social exclusion, rather than assimilation, wholly based on race (Glazer 1965). African Americans were given the most dangerous jobs, excluded from labor organizing, and consistently subjected to unfair housing practices (Segrue 1996). Today, Detroit is one of the most segregated areas in the nation as 83% of all Detroiters are African American, while the larger metropolitan area is largely Caucasian (United States Census 2010).

As quickly as Detroit attracted residents and businesses, the city lost population and economic power. Most recent census figures report that Wayne County has been losing population at a rapid rate, second only to Louisiana's Orleans Parish following the wake of Hurricane Katrina. By 2006, a city of almost 2 million people in 1950 had fallen to the 912,062 it has today (United States Census 2010). There are many reasons for the decrease in population, mainly stemming from "white flight" in the late 1960s, the race riots, or rebellions, of 1943 and 1967, and most importantly the decline of southeast Michigan's main economic force, the automotive industry. Michigan's population overall continues to suffer under the weight of slow economic growth. Currently, in the City of Detroit, 31.4% of all people and 27% of families are below the poverty level, while 20.5% are unemployed. After the 2009 economic downturn, unofficial numbers are reported at 40-50% unemployment. The other result is blight and abandoned property. The vacant lot numbers are estimated at 80,000 and taxable parcels with structures at 40,000 (The Kirwan Institute for the Study of Race and Ethnicity 2007). With few job prospects, increases in taxes and poor public transportation and municipal infrastructure, many find it difficult to thrive in their daily life. These figures demonstrate the dire situation many Detroiters find themselves in, forcing them to find creative ways to feed their families.

The city's plan to rebound from this hardship has situated itself in the downtown area, where big business, tourism, and high-end property development are courted. Development remains mainly focused on the downtown and midtown areas, and has yet to spread sufficiently throughout the residential neighborhoods (Bryant 2004). Local perceptions of the developments on the riverfront are mixed. Almost all Detroiters find the new opportunities a breath of fresh air in a city thought to be abandoned. They are grateful for the downtown cleanup and improved aesthetics, but many anglers of Detroit wonder why their most frequented neighborhood parks continue to be neglected. On their shores there is no clean-up of trash, no park maintenance, and no available restroom facilities. Many feel uncomfortable fishing on the newly restored riverfront near downtown. They recognize that other recreationalists are receiving preferred attention and that parks in the downriver suburbs are regularly cleaned and restored. The feeling is that the city has abandoned its poor residents to attract out-of-towners to spend money at the next car show or music event, while they wait for their fishing spots to be built upon by the next luxury lofts.

The Risks, the River, the State

Over the long period of industrial development in Detroit, the ecosystem has seen severe degradation followed by a period of recovery. In 1969, researchers discovered high levels of PCBs and mercury, two to four times higher than current recommended levels, in walleye when searching for DDT (Read at al. 2003). From this point the Ontario Water Resources Commission began monitoring mercury, finding that Dow Chemical of Canada was one major source of mercury (Read et al. 2003). The alarming news resulted in the complete shutdown of commercial fisheries from southern Lake Huron to western Lake Erie, costing an estimated $1 to 2 million dollars annually for local fisheries. Mercury-producing factories eliminated the contaminant in production in the early seventies. Since then, mercury levels have decreased in fish by 80% (Read et al. 2003). However, in 2009, the U.S. Geologic Survey found that 100% of U.S. fisheries contained mercury, perpetuated by coal-fired power plants emitting mercury into the water cycle (USGS 2009). Although PCBs were banned in 1979, many of the PCB-containing materials still in service during the ban were not required to be removed from use (EPA 1999).

As the issue of water contamination came to the forefront, so did the concern over accumulation of toxins in fish and their potential human health effects. Bioaccumulation and biomagnification are two crucial issues in the consumption of contaminated fish. Toxins accumulate in the bodies of fish, as bottom dwelling and smaller species are consumed by fish higher on the food chain. Researchers found that the longer the food chain, the larger the fish, and the closer a species is to urban-industrial centers, the higher the PCB or mercury level (Rasmussen et al. 1990; Cabana et al. 1994; Burreau et al. 2006). Regular consumption of Great Lakes fish increases the level of these contaminants in their consumers' bodies (Anderson et al. 1998).

The most prominent toxins found in waterways that accumulate in fish are mercury and PCBs. PCBs, used in industrial compounds, have been found to bioaccumulate in the fatty tissues of birds and fish since the late 1960s. The chemical, produced by Monsanto, was favored in a variety of industrial processes due to its level of fire resistance and low volatility until its banning in 1979 (Read et al. 2003). In 1972, researchers at the Canadian Wildlife Service found bioaccumulation of PCBs in eggs of herring gulls on Fighting Island in the Detroit River (Read et al. 2003). Additional studies have found that humans who consume large quantities of fatty-fish also tend to bioaccumulate PCBs in their body (Fängström et al. 2002). PCBs have been found to cause cancer through the study of workers exposed directly to the chemicals (Loomis et al. 1997). Other studies link PCB exposure to low birth weight, neurological damage and early developmental effects in children, ranging from motor skills to short-term memory (Agency for Toxic Substances and Disease Registry 2007). Women of childbearing age are particularly at risk due to the ability of PCBs to travel through breast milk and through the umbilical cord (Soechitram et al. 2004).

Mercury, on the other hand, has been known to enter the aquatic system through deposition, and rather than binding to fat like PCBs, is bound to muscular proteins (EPA 2001). Pregnant women exposed to mercury also increase the risk of brain damage to their fetus; mercury affects the growth of the baby's nervous system and brain, impairing development in language, memory, attention, and cognition (Myers and Davidson 1998; EPA 2007a). Mercury has been known to affect neurological development, especially in children, specifically through fish consumption, since the 1960s (Takeuchi et al. 1962). Much like the character of Alice in Wonderland's Mad Hatter, who is believed to have suffered from mercury poisoning, small doses of mercury can cause tremors, mood swings, twitching, and diminished cognitive function. Higher exposures may affect the kidneys and cardiovascular system, and may cause death in extreme cases (Nierenberg et al. 1998; Salonen et al. 2000).

Several other chemical contaminants are persistent in waterways, like dioxins, furans, pesticides, heavy metals, and others that are not currently monitored in most waterways or present in many fish advisories (EPA 2007b). Emerging issues of fire retardants, pharmaceuticals, and personal care product waste are also potential pollutants for consideration as fish contaminants. In 1988, the EPA reported that nearly three million cubic meters of municipal waste water was dumped in the Detroit River, matched by industrial effluent from power plants, steel mills, petroleum refineries, and chemical, automobile and plastics manufacturers. Much of the effluent on the U.S. side has been diverted through the Detroit Waste Water Treatment Plant since 1977, which now is the principal source of 15 contaminants including PCBs, cadmium, nickel, zinc, oil, cyanide, and ammonia (Manny and Kenaga 1991).

As a result of industrial pollution, several regulatory and research entities have taken charge of a range of aspects of testing and monitoring fish and wildlife, including human interactions with the Detroit River and industrial contamination. In 1987 the U.S. and Canada Great Lakes Water Quality Agreement spearheaded efforts to recover the Great Lakes region, creating the Great Lakes' Areas of Concern (AOCs). AOCs are defined as "geographic areas that fail to meet the general or specific objectives of the agreement where such failure has caused or is likely to cause impairment of beneficial use of the area's ability to support aquatic life." The U.S. and Canadian governments have identified 43 such areas: 26 in U.S. waters, 17 in Canadian waters, and five that are shared between U.S. and Canada on connecting river systems (Great Lakes Information Network 2008). Of the 14 beneficial use impairments, those for the Detroit River include: restrictions on fish and wildlife consumption, tainting of fish and wildlife flavor, restrictions on drinking water consumption, degradation of drinking water taste and odor, degradation of fish and wildlife populations, fish tumors or other deformities, degradation of aesthetics, and loss of fish and wildlife habitat (EPA 2007c). This agreement laid the groundwork for the Remedial Action Plan (RAP) to jointly assign responsibilities to recover and delist the Detroit River as an AOC. Detroit River RAP priorities include control of combined sewer overflows (CSOs), control of sanitary sewer overflows (SSOs), point/nonpoint source pollution controls, remediation of contaminated sediments, habitat restoration, and pollution prevention.

Communicating the Risks of Contaminated Fish

Since the mid-1970s, many states began looking at ways to protect their constituents from potential contamination by creating fish consumption advisories (Tilden et al. 1997). The advisories incorporate specific guidelines for people to safely eat fish that include size, species and number of meals for a given time period for each population, with more vulnerable populations typically receiving more stringent advice. State public health and/or environmental agencies create and issue fish consumption advisories in a wide variety of ways (Imm et al. 2005). Some are statewide advisories and others are smaller scale advisories on a county or watershed level, often depending on how their local governments are run and which agency is responsible for issuing the advisory. Further variation occurs as to the type of advisories. Advisories can be specific to either water bodies or regions and can apply only to commercially caught fish. The lack of a national mandate or regional guidelines for creating state specific advisories leads to confusion and extremes in the quality of the advisory and methods of outreach.

The number of fish advisories that are in effect has grown substantially since their inception. According to the EPA, the total number of advisories nationwide had grown to 3,852 by 2006. This amounts to a total of 38% of the nation's lakes, or 15,368,068 lake acres and 26% of total river miles, or 930,938 miles of river. All of the Great Lakes States of Illinois, Indiana, Michigan, Minnesota, New York, Ohio, Pennsylvania and Wisconsin include 100% of their lakes under a fish consumption advisory, and all but Minnesota and Michigan have 100% of rivers included. Michigan and Minnesota have only 3.5% of their rivers under a fish advisory (EPA 2007d). One of the Michigan rivers included under the advisory is the Detroit River.

The Michigan Department of Environmental Quality (MDEQ, which merged with the Michigan Department of Natural Resources in January 2010) sets trigger-levels, the total allowable level of contaminants, by testing lean dorsal fillets of several different species through the Fish

Contaminant Monitoring Project, in accordance with EPA and FDA recommended standards for mercury and PCBs respectively. The Michigan Department of Community Health (MDCH) is responsible for the creation and dissemination of that information as an advisory. However, due to state budget constraints, the advisory is available only on the DNR and MDCH websites. The actual fish consumption advisory is a chart that specifies how many of a particular species and size in a particular body of water are acceptable to eat per month. In Michigan, six species that live within the Detroit River are listed that have consumption limits (Canada recommends restricted consumption of nine species). The advisory considers the average meal to be 0.5 lbs. of fish and recommends that women and children, considered sensitive sub-populations, eat less fish per month than the average male angler weighing 155 lbs. The advisory's recommendations aid in making the decision to avoid potentially adverse effects of PCBs, mercury, and in some cases, dioxin. Yet, reading the Michigan guide is a navigation through circles, squares, triangles, little boxes and numbers. In handing over the guide to anglers, one must receive a mini-crash course in reading it (Fig. 2.1).

Legend:
- ▲ No eating restrictions
- ● One meal per month
- ▼ One meal per week
- ■ Six meals per year
- ◆ Do not eat these fish

Water body	Type of fish	Chemical(s)	General Population — Length (inches)									Women & Children — Length (inches)								
			6-8	8-10	10-12	12-14	14-18	18-22	22-26	26-30	30+	6-8	8-10	10-12	12-14	14-18	18-22	22-26	26-30	30+
Lake Erie Watershed (For water bodies that are not listed, read the Mercury Advisory on page 5)																				
Cass Lake* (Oakland Co.)	Smallmouth Bass	Mercury, PCBs				▼	▼	▼	▼							●	●	●	●	
	Walleye	Mercury, PCBs				▼	▼	▼	▼	▼					●	●	●	●	●	
Clear Spring Lake* (Macomb Co.)	Largemouth Bass	Mercury, PCBs				▲	▼	▼	▼							▼	●	●	●	
Clinton River (Downstream of Yates Dam, Macomb Co.)	Carp	PCBs	▲	▲	▲	▲	▲	▲	▲	▲	▲	●	●	●	●	●	■	■	■	■
	Rock Bass	PCBs	▲	▲	▲	▲	▲					▲	▼	▼	▼	▼				
	Suckers		▲	▲	▲	▲	▲	▲	▲	▲		▲	▲	▲	▲	▲	▲	▲	▲	▲
Detroit River	Carp	PCBs, Dioxins	◆	◆	◆	◆	◆	◆	◆	◆	◆	◆	◆	◆	◆	◆	◆	◆	◆	◆
	Freshwater Drum	Mercury, PCBs	▲	▲	▲	▲	▲	▼	▼	▼	▼	●	●	●	●	●	●	●	●	●
	Northern Pike	PCBs						▲	▲	▲								●	●	●
	Sturgeon	Mercury, PCBs	◆	◆	◆	◆	◆	◆	◆	◆	◆	◆	◆	◆	◆	◆	◆	◆	◆	◆
	Suckers	PCBs	▲	▲	▲	▲	▲	▲	▲	▲		▼	▼	▼	▼	▼	●	●	●	
	Walleye	PCBs				▲	▲	▲	▲	▲					●	●	●	●	●	●
	Yellow Perch	PCBs	▲	▲	▲	▲	▲					▲	▲	▼	▼	▼				
	All Other Species	PCBs, Dioxins, Mercury	▼	▼	▼	▼	▼	▼	▼	▼	▼	●	●	●	●	●	●	●	●	●
Ford Lake* (Washtenaw Co.)	Black Crappie	PCBs	▲	▲	▲	▲	▲	▲				▼	▼	▼	▼	▼	▼			
	Carp	PCBs	▲	▲	▲	▲	▲	▲	▲	▲	▲	●	●	●	●	●	●	●	●	●
	Channel Catfish	PCBs				▲	▲	▲	▲	▲	▲					▼	▼	●	●	●
	Walleye	PCBs					▲	▲	▲	▲	▲					▼	▼	▼	▼	▼
Hudson Lake* (Lenawee Co.)	Carp	Mercury	▲	▲	▲	▲	▲	▲	▲	▼	▼	▲	▲	▲	▲	▲	▲	▲	●	●
	Largemouth Bass	Mercury				▲	▼	▼	▼							▲	●	●	●	
Kent Lake* (Oakland Co.)	Black Crappie	Mercury, PCBs	▲	▲	▼	▼	▼	▼				▲	▲	●	●	●	●			
	Carp	PCBs	▲	▲	▲	▲	▲	▲	▲	▲	▲	●	●	●	●	●	●	●	●	●
	Largemouth and Smallmouth Bass	PCBs					▲	▲	▲	▲						▼	▼	▼	▼	
	Walleye	PCBs					▲	▲	▲	▲	▲					●	●	●	●	●

* See Mercury Advisory on page 5.

Fig. 2.1. 2009 Michigan Family Fish Consumption Guide (Michigan Department of Community Health 2009)

In the U.S. there is no uniform guide for fish consumption advisories. However, the Great Lakes states of Illinois, Indiana, Michigan, Minnesota, New York, Ohio, Pennsylvania and Wisconsin have written a Protocol for a Uniform Great Lakes Sport Fish Consumption Advisory (Anderson et al. 1993). Though most states have utilized parts of the protocol for the regional advisory, they adapted it for their own needs (Fischer et al. 1995). The Detroit River remains a special case in that its riparian areas are bi-national. Fish Consumption Advisories are also generated by Health Canada federally and communicated by the Ministry of Environment in Ontario. The Ministry of Environment tests fish, and develops and disseminates fish advisories to its anglers wholly separately from the U.S. so that anglers that fish from both the Detroit and Windsor side are receiving two different messages on the river. Since many Detroiters visit Canada to fish, they are aware, but confused by the discrepancy in advisories (Kalkirtz et al. 2008). For example, the Michigan advisory prohibits the consumption of carp due to PCBs and dioxins, yet Canada recommends that consumption of carp be permissible up to four times per month depending on the consumer's age and gender and the size of the fish (Fig. 2.2). The current status of the fish consumption advisory, though much needed and well intentioned, leaves cause for concern. The information provided by the fish consumption advisory is either not received through dissemination, or the message is not consistent and comprehensive for all populations.

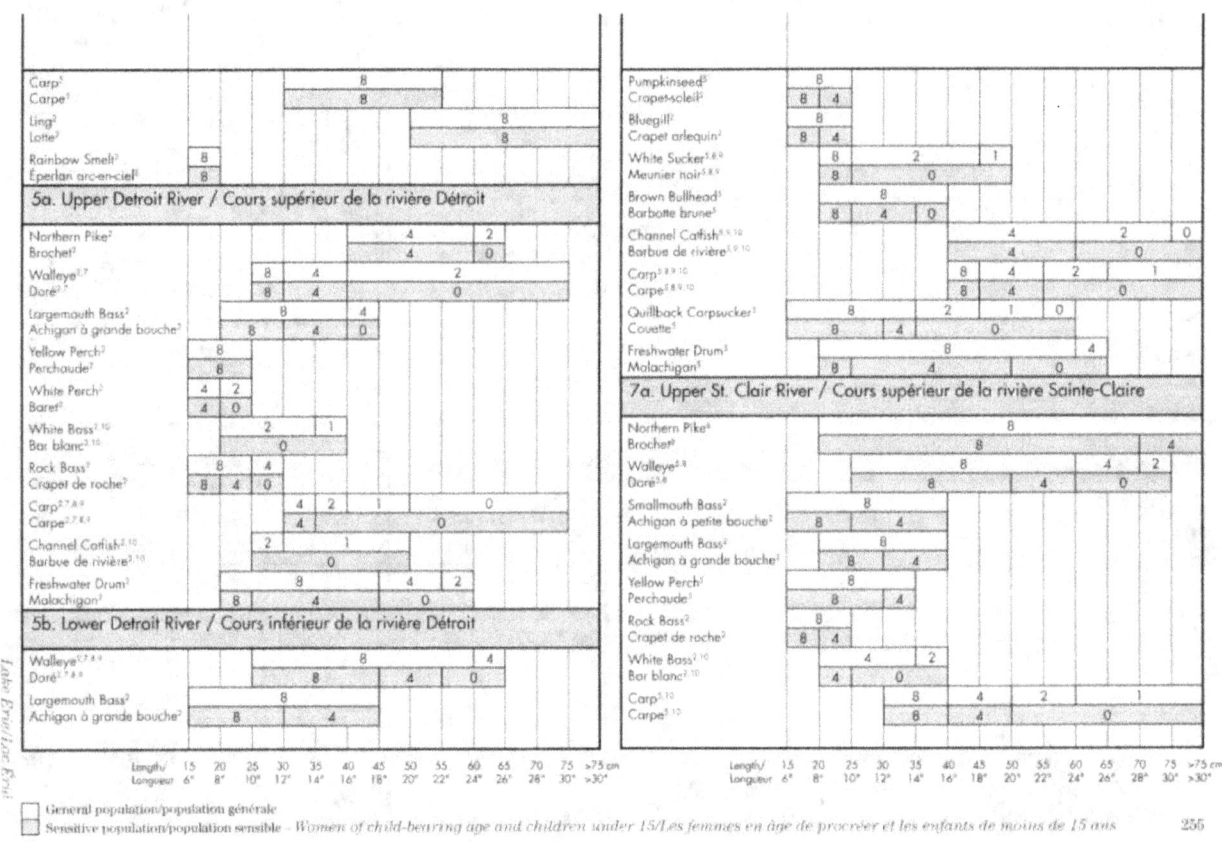

Fig. 2.2 2009-2010 Guide to Eating Ontario's Sport Fish (Ontario Ministry of the Environment 2009)

Environmental Justice and Fish Contamination Advisories

Environmental justice is the notion that justice must be sought for those who bear the unequal burden of pollution from industry (Bryant 2004). Statistically, low-income communities and communities of color are disproportionally burdened by pollution and environmental hazards (Bullard et al. 2007). Environmental justice relies heavily on the concepts of equity and social justice and is ecologically holistic as it incorporates every aspect of wildlife and human existence (Bryant 2004). The history of environmental justice is well documented. Events like Love Canal in Niagara Falls, New York, in 1978, in which local landfills were used as chemical waste dumpsites, brought the practice of toxic dumping and unequal burden to national attention (EPA 2009). Events such as the Warren County, North Carolina, community's protest against the development of a landfill "not only gained national attention, but led to subsequent studies and conferences on the differential exposure of environmental hazards on low-income and communities of color" (Bryant 2004). In 1987, shortly after the protest, the United Church of Christ of Ohio issued a report called Toxic Wastes and Race, which correlated industrial facility siting and pollution with race (Bullard et al. 2007). Subsequently, many studies have documented that, "[communities] of color have borne greater health and environmental risks than society at large" (Bullard 2007). Since the 1980s, more and more cases of industrial siting in neighborhoods of color—unequal industrial siting—have been documented throughout the U.S. Between 1989 and 1993 the number of people living in a zip code that contains both industry and hazardous waste facilities *increased* from 25% to 31% (Bryant 2004).

Currently in Detroit there are more than 80 industrial facilities that have reported hazardous waste activities, and more than 20 chemicals reported by the Environmental Protection Agency (EPA) that are toxic to the environment and human health (EPA 2006). Within the boundaries of the City of Detroit, people of color are continually exposed to contaminated land, air and water (SEMCG 2001; Keeler et al. 2002; Hartig and Stafford 2003). Several environmental justice and fish consumption studies specifically reference the disproportionate burden of contamination on people of low income and people of color. These studies provide the backdrop for a discussion on the anglers of the Detroit River and fish consumption advisories. In the City of Detroit this is an environmental justice issue.

Approximately half of all Great Lakes anglers report to have heard of their state fish consumption advisory and those numbers are tempered by race, sex, educational attainment, and consumption level (Tilden et al. 1997). Imm et al. (2005) reported that whites were more than six times more likely to be aware of their state's advisory than blacks, and men were four times more likely to be aware than women. This is a reflection of a national trend that information distributed by the state may not reach all of the population equally. Burger et al. (1999) find that people of color and non-English speaking immigrants are less aware of fish advisories or the health risks from eating fish. This exacerbates the risk of eating fish for different ethnic groups that prefer fish as their source of protein. Another study by Burger and Gochfeld (2006) found that while patterns of fish consumption vary ethnically, minorities and Native Americans consume more fish than whites. Not only are more people of color fishing from contaminated urban bodies of water at a higher percentage, people of color make up the predominate group that do not have access to or awareness of such information.

Why is the Message Not Reaching Anglers?

On the Detroit River, there are many reasons why anglers are not receiving the fish consumption advisory message. In many cases, even if the message reaches them, it is disregarded. Some of the issues include lack of access, difficulties in interpreting the complicated and inconsistent messages, and distrust of both the messenger and the message. Instead, anglers often find their own methods of determining the safety of eating fish caught from the Detroit River (Kalkirtz et al. 2008).

Many anglers lack access to the information designed to help protect them from unhealthy consumption patterns. In the State of Michigan, the 2009 Guide to Eating Sport Fish can be found only on the internet. Due to significant budget constraints, the advisory is no longer available in printed form, so anglers must be able to access the internet version for all of the advisory information. The lack of a physical advisory adds additional confusion and often leaves out the most vulnerable populations such as those of low income, people of color and recent immigrants with different cultural and dietary needs.

Additionally, a portion of the population who do not fish themselves receives fish from the Detroit River—those might include family or community members without the resources or knowledge to fish themselves. There is a question as to whether this population, fish consumers once removed from the act of fishing, is receiving adequate information to protect themselves. The estimated numbers of all Great Lakes anglers suggests that those who attain licenses represent less than half of those who consume sport fish, presenting a further complication for risk communication (Tilden et al. 1997). Those without a license would not have the opportunity to receive one at the point-of-purchase.

The Michigan and Ontario fish consumption guides often have conflicting information that is difficult for anglers to interpret. Common names of fish are not represented on the consumption advisory, nor are there pictures of fish. Catfish, a popularly caught fish for consumption, is not listed on the Michigan advisory, but is found on the Canadian. Steelhead, called such for its steel-like scales, is not present on the advisory. This fish, which is commonly regarded as "dirty" and full of contaminants is rather listed under "drum." In addition to the complicated nature and unavailability of the fish advisories, the message is inconsistent and not targeted to the anglers that need it most. It has not been a priority for Michigan's regulatory agencies to reach all anglers for reasons that are not entirely known or documented.

Fish consumption advisories are the only method by which anglers can receive information about contaminants within a local, specific body of water. In some cases, advisories are often explicitly ignored by the general public (Burger and Gochfeld 2006). Chess et al. (2005) find that government agencies fail to communicate effectively, using what they call "government speak." Institutionally generated information is often inadequate for minority populations who generally understand standard risk communication as culturally irrelevant. Often times the lack of meaningful input from community members creates distrust of the message. In 1994 a group of University of Michigan graduate students found this to be true on the Detroit River. They found that African American anglers did not trust the state or information generated by the state, skeptical of efforts to control angler behavior and not industrial pollution (Hornbarger et al. 1994). They also found that anglers in Detroit are also consuming a wider breadth of fish species more often, placing them in a higher risk category. Distrust of the government plays a role in the willingness of anglers to heed the advisories.

Instead of relying on state distributed consumption advisories, selection of fish for consumption is a personal decision, based on a personally crafted assessment of risk using one's own avoidance strategy, based on common knowledge (Beehler 2001). Common-based knowledge can be linked to fishing as a cultural activity. Fishing plays a critical role in the cultural fabric threaded along the river and throughout the Detroit metropolitan area. Much of the fishing skill acquired came from previous generations, perhaps brought from the South. This cultural knowledge is shared with children, nieces and nephews, not only making this an intergenerational issue, but also one of historical significance. More often than not it is reported that along the river, fishermen and women engage in a gift culture. Fishing is free. If you don't have the right gear, it will be shared, and if you catch a fish you will not eat, it is rarely thrown back before being offered to another angler. The gift culture of fish is important to African Americans on the Detroit River, and signifies a strong social capital with catching and sharing fish (Hornbarger et al. 1994). It also extends beyond the shores, where some anglers who do not eat fish, or catch beyond their consumption ability take the fish and give it to neighbors or elderly or cook it at a fish fry, inviting any and all. Fishing links communities together and has created healthy, safe, and virtuous traditions in Detroit.

For the anglers of the Detroit River, it is rarely stated that one depends on the resource for food. Rather a more common statement is that one fishes for leisure, pleasure, or enjoyment (Dawson 1999). Yet it is more commonly found that U.S. residents, as opposed to their Canadian counterparts, rely on the Detroit River for consumption rather than sport (Kalkirtz et al. 2008). Many people continue to fish, not due to preference, but rather due to "economic hardship" (Burger and Gochfeld 2006). This creates an urgent need to disseminate information to this population in a culturally relevant and easily understood manner. Self-caught fish provide a low-cost supplement to diet as store-bought fish is expensive and its place of origin unknown (Burger and Gochfeld 2006). This is closely related to the issue of food security in Detroit. With the recent merging of one of the few large grocers within the city limits, there are a total of five grocery stores over the size of 20,000 square feet, giving Detroit the designation of a "food desert"(Smith 2007). Detroit covers a large geographic area of 139 square miles, making it difficult for those who do not have vehicles to access grocery stores. And with notoriously poor public transportation options, residents without cars, or who cannot afford gas, are forced to rely on corner stores with higher prices and with little to no selections of fresh foods. A growing grassroots movement of community gardens has created more options for some residents, and many turn to fishing on the Detroit River to supplement their diets and those of their households and community. The question remains for anglers: Why would the state insist on limiting yet another source of food? Many Detroit residents, unlike their wealthy counterparts, cannot drive to a grocery store for more expensive sea bass or salmon. Low-income Detroit anglers cannot take the boat into Lake Huron or inland rivers where there are more species and better catch. For these reasons, anglers have voiced that the fish consumption advisory is seen as a source of frustration, or as a set of lies created by the state, another reason to limit behavior of anglers but not of industry, or just the same old song, not worth paying attention to because no one is sick.

The way in which the fish consumption guide is written and the manner by which fish are tested brings up other crucial questions. Are the anglers who reside on the Detroit River shores given the tools and information to adequately navigate such wicked questions as: Is eating fish bad or good for my health? What am I sacrificing if I cannot eat fish? Potential questions that would ensure adequate information and tools for anglers include: What is the literacy rate of subsistence anglers and what is the comprehension level needed for understanding these advisories and do those two things match up? Are advisories in Spanish and Arabic readily available to anglers at bait shops, as a large concentration of Latino and Arab Americans lives along the shores? Are all the species that anglers eat on the list of species tested by the state? Are they referred to as the same thing by those fishing on the shore? Is the state prepared to answer the questions of a people traditionally disenfranchised from the government, low-income African Americans, Latinos, Arab Americans and Caucasians, who use the Detroit River as a resource, daily, in a manner they understand without threatening their livelihood?

Rethinking the Scientific Approach: Making Science Readable and Speakable

Science can make us better activists. Activism can make science useful. In this case, science has done an amazing job in understanding the ways in which PCBs and mercury have resided in the Detroit River and how they have affected the food chain. Around the world scientists are delving deeper into the adverse effects of fish contaminants on the human body and its development. Yet it is the way in which science is communicated, or undercommunicated, to those most at risk that has become an issue. Fishing is of great value to all anglers on the Great Lakes, there is passion and excitement behind every catch. On the Detroit River shores there are barbeques and fish fries where everyone is invited. There are blue skies and carefree days, alleviation from the pressures of city life. There is a priceless value of access to and knowledge of the River, this in essence becomes a wager against the knowledge of consumption-limiting toxins.

Activists and advocates within government and non-government agencies are doing great things to provide crucial information to at-risk populations. Across the nation there are models by which we can craft new forms of governance. For example, campaigns that specifically target subsistence anglers for voluntary biomonitoring, where blood-levels can be tested for mercury levels upon inquiry, can aid in reducing fear and uncertainty around consumption. Another suggestion from Lori Verbrugge, Toxicologist and State Environmental Public Health Program Manager for Alaska, encourages a highly localized risk benefit analysis that includes nutritional, cultural, fitness, intergenerational, and ecological benefits while considering risks such as costs, contaminants, or obesity and heart disease due to a low fish diet (Verbrugge 2007).

Gary Ginsberg of the Connecticut Department of Public Health suggests that beyond just the "what" of risk communication, there should be a focus on the "how" of risk communication. Within that "how," agencies should include a focus on individual fish that promotes "good" fish, creating positive attention towards those safe to eat. This will take attention from a sanctioning role of the state, to a promotional. For example, recent developments as a result of research contributed by these authors, have impacted outreach efforts to include some new avenues such as adding signs to popular fishing spots and recruiting the assistance of local environmental justice organizations. This year, the MDCH partnered with Detroiters Working for Environmental Justice to work with urban youth to develop an outreach strategy utilizing interviews and alternative media on the relative risks of fish consumption.

More aggressive initiatives include the incorporation of science and environmental stewardship through educational models. Creating political allies within one's own community is crucial in following through on developing practices that represent good science. Policy-makers should come forward to offer grants, funding, and support programs that will close the gap between resource users and policy-makers. In an era of budget deficits, creative solutions must include collaborative partnerships between state and non-governmental actors, and between scientists and community activists that speak beyond institutional and epistemological divisions. We must not only learn to read new languages in differing fields, but speak new languages across class and racial boundaries. Under the umbrella of sustainable development and job security, we must continue to push for a clean environment for all members of our community that takes into account historical inequity to develop stronger bonds of stewardship that will carry us into the future.

Acknowledgements

The authors would like to thank Dr. Bunyan Bryant and Dr. Elaine Hockman for their remarkable support, guidance and dedication to the environmental justice community. We would also like to thank Dr. Donna Kashian and Dr. Larissa Sano of the Cooperative Institute for Limnology and Ecosystem Research (CILER) and Michigan SeaGrant for their assistance. This work has been made possible through the generous support of School of Natural Resources and Environment, Rackham Graduate School and the Environmental Justice Initiative at the University of Michigan in Ann Arbor.

References

Agency for Toxic Substances and Disease Registry, 2007. ToxFAQs™ for Polychlorinated Biphenyls (PCBs) (Bifenilos Policlorados (BPCs)). Available online at http://www.atsdr.cdc.gov/tfacts17.html.

Anderson, H., J. F. Amrhein, P. Shubat, and J. Hesse, 1993. Protocol for a Uniform Great Lakes Sport Fish Consumption Advisory, Great Lakes Fish Advisory Task Force Protocol Drafting Committee.

Anderson, H. A., C. Falk, L. Hanrahan, J. Olson, V. W. Burse, L. Needham, D. Paschel,

D. Patterson, R. H. Hill, and The Great Lakes Consortium, 1998. Profiles of Great Lakes critical pollutants: A sentinel analysis of human blood and urine. Environmental Health Perspectives, v. 105, pp. 279-289.

Beehler, G. P., B. McGuiness, and J. E. Vena, 2001. Polluted fish, sources of knowledge, and the perception of risk: Contextualizing African American anglers' sport fishing practices. Human Organization, v. 60, pp. 288-297.

Bryant, B., 2004. Environmental Advocacy: Working for Economic and Environmental Justice, 2nd ed. Bunyan Bryant, Ann Arbor, MI.

Bullard, R. D., 2007. Smart growth meets environmental justice. In Growing Smarter: Achieving Livable Communities, Environmental Justice, and Regional Equity, R. D. Bullard (ed.). The MIT Press, Cambridge, MA, pp. 23-50.

Bullard, R. D., P. Mohai, R. Saha, and B. Wright, 2007. Toxic Waste and Race at Twenty 1987- 2007: Grassroots Struggle to Dismantle Environmental Racism in the United States. United Church of Christ.

Burger, J., K. K. Pflugh, L. Lurig, L. A. Von Hagen, and S. Von Hagen, 1999. Fishing in urban New Jersey: Ethnicity affects information sources, perception, and compliance. Risk Analysis, v. 19, pp. 217-229.

Burger, J., and M. Gochfeld, 2006. A framework and information needs for the management of the risks from consumption of self-caught fish. Environmental Research, v. 101, pp. 275-285.

Burreau, S., Y. Zebuhr, D. Broman, and R. Ishaq, 2006. Biomagnification of PBDEs and PCBs in food webs from the Baltic Sea and the northern Atlantic Ocean. Science of the Total Environment, v. 366, pp. 659-672.

Cabana, G., A. Tremblay, J. Kalff., and J. B. Rasmussen, 1994. Pelagic food chain structure in Ontario Lakes: A determinant of mercury levels in lake trout (Salvelinus namaycush). Canadian Journal of Fish Aquatic Science, v. 51, pp. 381-387.

Chess, C., J. Burger, and M. H. McDermott, 2005. Speaking like a state: Environmental justice and fish consumption advisories. Society and Natural Resources, v. 18, pp. 267-278.

Dawson, J., 1999. Hook, Line, and Sinker: A Profile of Shoreline Fishing and Fish Consumption in the Detroit River Area. Fish and Wildlife Nutrition Project, Project No. K341813, Health Canada, Ottawa, ON.

Dunbar, W. F., 1965. Michigan: A History of the Wolverine State. W. B. Eerdmans Publishing Co., Grand Rapids, MI.

EPA (U.S. Environmental Protection Agency), 1999. Polychlorinated Biphenyls Update: Impact on Fish Advisories. Environmental Protection Agency, Office of Water, EPA-823-F-99-019.

EPA (U.S. Environmental Protection Agency), 2001. Mercury Update: Impact on Fish Advisories. Environmental Protection Agency, Office of Water, EPA-823-F-01-011.

EPA (U.S. Environmental Protection Agency), 2006. EPA Envirofacts. Available online at http://oaspub.epa.gov/enviro/ef_home3.html.

EPA (U.S. Environmental Protection Agency), 2007a. Mercury. Available online at http://www.epa.gov/mercury/effects.htm.

EPA (U.S. Environmental Protection Agency), 2007b. Laws and Regulations. Available online at http://www.epa.gov/lawsregs/.

EPA (U.S. Environmental Protection Agency), 2007c. Detroit River Area of Concern. Available online at http://epa.gov/greatlakes/aoc/detroit.html.

EPA (U.S. Environmental Protection Agency), 2007d. EPA Fact Sheet: 2005/2006 National Listing of Fish Advisories.

EPA (U.S. Environmental Protection Agency), 2009. EPA press release - December 20, 1979: U.S. Sues Hooker Chemical at Niagara Falls, New York. Available online at http://www.epa.gov/history/topics/lovecanal/02.htm.

Fängström, B., M. Athanasiadou, P. Grandjean, P. Weihe, and A. Bergman, 2002. Hydroxylated PCB metabolites and PCBs in serum from pregnant Faroese women. Environmental Health Perspectives, v. 110, pp. 895-899.

Fischer, L. J., P. M. Bolger, G. P. Carlson, J. L. Jacobson, B. A. Knut, M. J. Radik, M. A. Roberts, and P. T. Thoma, 1995. Critical Review of a Proposed Uniform Great Lakes Fish Advisory Protocol. Michigan Environmental Science Board, Lansing, MI.

Glazer, S., 1965. Detroit: A Study in Urban Development. Bookman Associates, New York.

Great Lakes Information Network, 2008. Areas of Concern (AOCs) in the Great Lakes Region. Available online at http://www.great-lakes.net/envt/pollution/aoc.html.

Hartig, J. H. and T. Stafford, 2003. The public outcry over oil pollution of the Detroit River. *In* Honoring Our Detroit River: Caring for Our Home, J. H. Hartig (ed.). Cranbrook Institute of Science, Bloomfield Hills, MI, pp. 69-78.

Hornbarger, K., C. MacFarlene, and C. R. Pompa, 1994. Target Audience Analysis: Recommendations for Effectively Communicating Toxic Fish Consumption Advisories to Anglers on the Detroit River. Technical Report #11, Natural Resource Sociology Research Lab, University of Michigan, Ann Arbor, MI.

Hyde, C., 1980. Detroit: An Industrial History Guide. Ninth Annual Conference, Society for Industrial Archeology, Detroit, MI.

Imm, P., L. Knobeloch, H. A. Anderson, and the Great Lakes Sport Fish Consortium, 2005. Fish consumption and advisory awareness in the Great Lakes Basin. Environmental Health Perspectives, v. 113, pp. 1325-1329.

Kalkirtz, V., M. Martinez, and A. Teague, 2008. Environmental Justice and Fish Consumption on the Detroit River Area of Concern. Master's Project, School of Natural Resources and Environment, University of Michigan, Ann Arbor, MI.

Keeler, G. J, J. T. Dvonch, F. Y. Yip, E. A. Parker, B. A. Israel, F. J. Marsik, M. Morishita, J. A. Barres, T. G. Robins, W. Brakerfield-Caldwell, and M. Sam, 2002. Assessment of personal and community-level exposures to particulate matter among children with asthma in Detroit, Michigan as part of community action against asthma (CAAA). Environmental Health Perspectives, v. 110, Suppl. 2, pp. 173-181.

Kerr, J., S. Olinek, and J. H. Hartig, 2003. The Detroit River as an artery of trade and commerce. *In* Honoring Our Detroit River: Caring for Our Home, J. H. Hartig (ed.). Cranbrook Institute of Science, Bloomfield Hills, MI, pp. 35-48.

The Kirwan Institute for the Study of Race and Ethnicity, 2007. Land banking in Detroit. Available online at http://kirwaninstitute.org/news/news_landbankdetroit.html.

Loomis, D., S. R. Browning, A. P. Schneck, and D. A. Savitz, 1997. Cancer mortality among electric utility workers exposed to polychlorinated biphenyls. Occupational and Environmental Medicine, v. 54, pp. 720-728.

Manny, B. A. and D. Kenaga, 1991. The Detroit River: Effects of contaminants and human activities on aquatic plants and animals and their habitats. Hydrobiologia, v. 219, pp. 269-279.

Michigan Department of Community Health, 2009. Michigan Family Fish Consumption Guide. Available online at http://www.michigan.gov/dnr/0,1607,7-153-10364---,00.html.

Myers, G. L. and P. W. Davidson, 1998. Prenatal methylmercury exposure and children: Neurologic, developmental, and behavioral research. Environmental Health Perspectives, v. 106, Suppl. 3, pp. 841-847.

Nierenberg, D. W., R. E. Nordgren, M. B. Chang, W. Siegler, M. G. Blayney, F. Hochberg, T. Y. Toribara, E. Cernichiari, and T. Clarkson, 1998. Delayed cerebellar disease and death after accidental exposure to dimethyl mercury. New England Journal of Medicine, v. 338, pp. 1672-1675.

Ontario Ministry of the Environment, 2009. Guide to Eating Ontario Sport Fish, 24th ed. Available online at http://www.ene.gov.on.ca/envision/guide.

Rasmussen, J. B., D. J. Rowan, D. R. S. Lean, and J. H. Carey, 1990. Food chain structure in Ontario lakes determines PCB levels in Lake Trout (*Salvelinus namaycush*) and other pelagic fish. Canadian Journal of Fisheries and Aquatic Sciences, v. 47, pp. 2030-2038.

Salonen, J. T., K. Seppanen, T. A. Lakka, R. Salonen, and G.A. Kaplan, 2000. Mercury accumulation and accelerated progression of carotid atherosclerosis: a population-based prospective 4-year follow-up study in men in eastern Finland. Atherosclerosis, v. 148, pp. 265-273.

Read, J., D. Haffner, and P. Murray, 2003. Mercury and PCB contamination of the Detroit River. *In* Honoring Our Detroit River: Caring for Our Home, J. H. Hartig (ed.). Cranbrook Institute of Science, Bloomfield Hills, MI, pp. 91-106.

Salonen, J. T., K. Seppänen, T. A. Lakka, R. Salonen, and G. A. Kaplan, 2000. Mercury accumulation and accelerated progression of carotid atherosclerosis: A population-based prospective 4-year follow-up study in men in Eastern Finland. Atherosclerosis, v. 148, pp. 265–273.

Segrue, T., 1996. The Origins of Urban Crisis. Princeton University Press: Princeton, NJ.

Sidhu, K., 2003. Health benefits and potential risks related to consumption of fish or fish oil. Regulatory Toxicology and Pharmacology, v. 38, pp. 336-344.

Smith, J. J., 2007. Grocery closings hit Detroit hard. The Detroit News, July 5, 2007.

Soechitram, S., M. Athanasiadou, L. Havander, A. Bergman, and P. J. J. Sauer, 2004. Fetal exposure to PCBs and their hydroxylated metabolites in a Dutch cohort. Environmental Health Perspectives, v. 112, pp. 1208-1212.

SEMCG (Southeast Michigan Council of Governments), 2001. Water Quality Issues in Southeast Michigan. Detroit, MI.

Takeuchi, T., N. Morikawa, H. Matsumoto, and A. Shiraishi, 1962. A pathological study of Minamata disease. Acta Neuropathologica (Berlin), v. 2, pp. 40-57.

Tilden, J., L. P. Hanrahan, C. Palit, J. Olson, W. MacKenzie, and The Great Lakes Fish Consortium, 1997. Health advisories for consumers of Great Lakes sport fish: Is the message being received? Environmental Health Perspectives, v. 105, pp. 1360-1365.

United States Census, 2010. Available online at http://www.census.gov.

USGS (United States Geological Survey), 2009. National Water Quality Assessment Program. Available online at http://water.usgs.gov/nawqa/mercury/.

Verbrugge, L., 2007. Fish consumption advice goes local. EPA Forum on Contaminants in Fish. Portland, ME.

Woodford, A. M., 2001. This is Detroit 1701-2001. Wayne State University Press, Detroit, MI.

Zunz, O., 2000. The Changing Face of Inequality: Urbanization, Industrial Development, and Immigrants in Detroit, 1880-1920. University of Chicago Press, Chicago, IL.

Chapter 3: Bisphenol A: Mothers Shouldn't have to be Scientists

John Dao and Rebecca Roberts

Summary

Bisphenol A (BPA) is a polymerizing agent used to make polycarbonate plastic and epoxy linings found in some water bottles, baby bottles, food storage containers, food and beverage cans, tableware, toys, and some medical devices. BPA ranks among the highest volume chemicals manufactured worldwide. Exposure to BPA affects normal development and health by altering estrogen signaling in the hormonal system. Over 92% of Americans have detectable levels of BPA in their bodies, with infants, children, and the poor bearing the highest burden of BPA exposure. Prenatal exposure is especially detrimental. Animal studies show that prenatal BPA exposure causes behavioral abnormalities, malformed urethra and genital tracts, higher prostate cancer risk, and, in females, earlier puberty. Studies link BPA to increased risk for obesity, heart disease, and diabetes. Pregnant women and mothers must be informed as to the importance of reducing their BPA exposure, and thus the exposure of their unborn babies and children. Yet, they often do not have the information needed to make the best lifestyle and consumer decisions. Information is available, but the history of BPA safety has been fraught with battles between stakeholders, such as the scientists and the chemical industry, and has led to confusion and mixed messages. Although no federal regulatory agency has moved to limit exposure to BPA and the chemical industry continues to insist it is safe, the Health and Human Services (HHS) web site, for the first time, in 2010, posted some useful tips for parents on how to limit exposure. This presentation of mixed messages—it is "safe" but here's how to avoid exposure—challenges the ordinary mother, who shouldn't have to be a scientist to decipher the BPA controversy and to decide a course of action regarding use of BPA products.

The Roles of Scientists and Mothers

My son loves his baby bottle. (First-person narrative is from the viewpoint of Rebecca Roberts.) Some kids have teddy bears or blankies, my son has his ba-ba. But as I sit and watch him play trains with a bottle hanging limply from his lips, I cringe a little bit. I wonder how many other babies and kids are drinking from bottles or sippy cups right now.

Perhaps an odd thing to ponder, but unlike most mothers, I'm also a scientist. I'm a scientist who studies plastic. More specifically, I study a compound in some plastics called bisphenol A. This compound is part of the plastic that makes up some baby bottles. I also know that scientific research shows that exposure to bisphenol A, especially in babies, is detrimental to good health. I cringe because I know this, but most mothers do not. Most mothers don't realize that they may be exposing their children to toxic chemicals present in that baby bottle. I've purposefully purchased bisphenol A-free bottles for my son. I wonder how many other mothers have followed in my footsteps and thrown away delightfully-decorated baby bottles received at their baby shower in an attempt to reduce the level of bisphenol A their child is exposed to?

How do mothers even know that they should be cautious about exposing their babies to bisphenol A? How many of them, when learning of the wonderful news that they were pregnant, knew that they should minimize their own exposure to bisphenol A to protect their growing fetus? Does every mother need to be a scientist in order to know this information and protect herself and her children?

To answer this question, I would like to first discuss what it means to be a scientist and a mother. As someone who carries out both roles, I appreciate that the priorities for each are different.

As a scientist I depend on evidence, logic, and imagination to explain observations that I make in the laboratory. My role is to generate and interpret scientific data in a nonbiased and objective way. I also convey the data to the scientific community and the public. Moreover, I need to interpret the data of other scientists and need to be aware of any lack of objectivity in their work. My social responsibility as a scientist is to produce nonbiased evidence and share my knowledge in an impartial manner. Thus, there are two aspects of my life as a scientist— generating information and communicating information.

As a mother, my role is to raise a healthy and happy child. To do this, I must make daily decisions as to what my child does and does not do. I must choose what he is exposed to and try to limit danger in his life. I must have information in order to make these decisions.

So, when it comes down to it, what I really depend on in both of my roles is information—non-biased, factual, evidence-based information. I am both a producer and a consumer of information. For some issues, the mother in me relies on my training as a scientist to objectively look at scientific information and determine the personal choice for my son. But, like all people, I am not qualified, nor do I have the time, to understand all scientific issues. At this point, the mother must rely on others in society—on the brokers of information.

Who Are The Scientific Information Brokers?

Who are these brokers of information? They are other scientists, medical personnel, the government, regulatory agencies, corporations, non-profit organizations, and the media, to name a few. I rely on all of these groups to assist me in making informed decisions about my life and the life of my child. As I reflect on my role as a mother and my dependency on others to assist my decision-making process and, in reality, often make decisions for me, I realize that I should not take their decisions at face value. That the purveyors of information are not necessarily as objective in their interpretation and dissemination of scientific data as my scientific self would like. That behind every decision is timing, money, convenience, politics, and countless other agendas.

As a person with dual roles on either side of the information spectrum, I certainly have an appreciation for the not-so-straight path along which information is communicated, but I still must rely on others to assist me. Unlike the majority of mothers, I might better understand the science behind bisphenol A and its effects on human health simply because I work within that scientific world, but I still rely on the scientific results of others to comprehend the topic more fully.

The story of bisphenol A is only one small issue in the larger picture of science and human health. Take smoking for example. As a mother I will do everything I can so that my son doesn't smoke. Why? Because I know that it is unhealthy and dangerous. But how do I know? I haven't actually seen the data. I haven't performed the experiments. I suppose my scientist self could gather the information that's available and learn the topic, interpret it for myself. But life is busy. I must rely on others. I rely on society, the media, the American Cancer Society, the Surgeon General to tell me that it's dangerous to smoke. They set the boundaries for me both by telling me what they believe I need to know and by setting governmental limits on tobacco use (for example by banning the sale of tobacco products to children under the age of 18). So, I suppose this is where my problem of balancing my role as scientist with my role of mother begins. I can do my best to provide objective scientific data to the brokers of information. But as every other mother, I need to rely on others in society to provide information, interpret data, set limits, and assist me in making the daily decisions involved in raising my son. I can only hope that the information I receive is objective and the boundaries and regulations set are for the benefit of my family and society as a whole.

This is what it comes down to—mothers are not scientists. They shouldn't have to be in order to raise healthy children. But all mothers, and all parents, need clear and correct information in order to make healthy decisions for their children. And beyond this, all of us, parent or not, need information to make healthy decisions for ourselves.

So, how is the information regarding the effect of bisphenol A on human health being packaged and communicated to the general public? (And I must now point out that in writing this essay I have breached the comfort zone of my previous roles of a mere producer and consumer of information and have trespassed into the realm of the information broker. As such, I am no longer simply producing, collecting, interpreting, and reporting the available data, but am also conveying a personal value assessment on the information.)

The Bisphenol A in our Environment

Let's begin by understanding what bisphenol A is and how our modern society relies on it.

In 1952 polycarbonate plastics were first generated by chemists using the chemical bisphenol A, or BPA. These plastics are used to make such products as compact discs, sunglasses, bicycle helmets, water and milk bottles, baby bottles, food storage containers, tableware, plastic windows, bullet resistant laminate, cell phones, car parts, toys, and some medical devices such as incubators, dialysis machines, and blood oxygenators. BPA is also used to make certain resins that are commonly found in the linings of food cans to prevent corrosion. Moreover, BPA is present in some polyvinyl chloride (PVC) plastic products, in white dental fillings, dental sealants, and in some flame retardants. Recent findings show that BPA is also used in the manufacture of some thermal papers, the heat-sensitive paper used for cash register receipts (Biedermann et al. 2010; Mendum et al. 2010). Clearly, many useful and essential products used in our society contain BPA. Accordingly, it is one of the chemicals produced in the highest volume worldwide, with an annual production in 2003 of six billion pounds (Lyons 2000; Vom Saal 2005). The production of BPA in the United States alone generates about six billion dollars a year for the chemical industry (Borrell 2010). Regulation requiring a significant reduction in the production and use of BPA could have a drastic economic impact, not to mention the daily changes in personal lifestyle that would follow.

The concerns regarding BPA come about because BPA has been shown to leach from products that contain it, such as water bottles and food cans, into the foodstuffs stored in the container (Brede et al. 2003; Goodson et al. 2004; Munguia-Lopez et al. 2005). When the foodstuffs are then consumed, BPA enters the body through the digestive tract. BPA is also present in rivers and streams and in drinking water, presumably due to leaching from plastic items in landfills (Kuch and Ballschmiter 2001; Kolpin et al. 2002; Coors et al. 2003; Loos et al. et al. 2003; Boyd et al. 2004; Jackson and Sutton 2008). Considering the number of products that are manufactured with BPA, human exposure to this compound occurs in many ways. However, ingestion of food or liquid that was stored in BPA-containing plastic (this includes BPA-lined food cans) is a significant route of exposure.

The level of BPA released from plastic is dependent on the age and wear of the plastic and exposure to heat. For example, one study showed that small levels of BPA were leached from baby bottles that were subjected to simulated use that included boiling, washing with a bottle brush, and dishwashing (Brede et al. 2003). Plastic tableware (such as those used in some schools) was also found to release BPA into hot vegetable soup (Lyons 2000; Brede et al. 2003). Older, worn bottles and bowls released BPA more readily than newer products (Lyons 2000; Brede et al. 2003).

BPA is also found in the inner linings of metal food cans (Goodson et al. 2004; Munguia-Lopez et al. 2005). The epoxy lining protects the metal can from rust and corrosion. While damage and age of the cans does not appear to greatly increase the amount of BPA leached into the food, the initial packaging and sterilization techniques do release significant levels of BPA into the food (Goodson et al. 2004; Vandenberg et al. 2007). Indeed, consumption of canned food is thought to be a major route of BPA exposure.

Who Carries the BPA Burden?

A survey by the Centers for Disease Control found that approximately 92% of Americans have detectable levels of BPA in their bodies (Calafat et al. 2005, 2007). The survey is considered representative of the United States population even though it included only people older than six years old. It is notable, however, that the children (ages 6-11 years) in this study displayed the highest levels of all populations investigated. A small study in 2009 indicated that nine out of ten samples of newborn cord blood contained BPA (EWG 2009). BPA has been found in placental tissue and fetal blood as well, indicating that fetuses are being exposed as a result of maternal exposure (Schonfelder et al. 2002).

An additional study looked at the levels of BPA in infants—specifically in premature infants housed in Neonatal Intensive Care Units (Calafat et al. 2009). It was found that this population displayed the highest level of BPA in their bodies. The primary route of exposure for these infants appears to be due to the use of invasive, BPA-containing medical devices and possibly due to the use of premade liquid formula that is packaged in BPA-containing bottles (Kuo and Ding 2004; Green et al. 2005; Houlihan 2007). Of note, BPA can be passed to the infant through breast milk as well, but the overall benefits of breastfeeding should be considered before avoiding breast milk due to BPA contamination (Sun et al. 2004; Houlihan 2007; Kuruto-Niwa et al. 2007).

It is clear that infants and children are burdened with the highest levels of BPA. This is likely due to several factors. First, the ability of infants and children to metabolically detoxify contaminants such as BPA is not yet mature. While BPA will ultimately be removed from the blood through a liver-mediated process, BPA stays in the system of infants and children longer than in adults (Vom Saal et al. 2007). Second, when compared to adults, infants and children consume proportionately more food when their overall body size is taken into account and therefore bodily concentrations are higher. The milk, formula, or food that is the main staple of an infant is often stored in containers (baby bottles, sippy cups) that are made with BPA (Kuo and Ding 2004; Calafat et al. 2007; Vandenberg et al. 2007). Infants and young children are also more likely to mouth plastic products not necessarily meant to be put in the mouth, such as toys or the hard plastic portion of a pacifier. Finally, BPA can also pass from the mother to the fetus (Schonfelder et al. 2002).

Given these findings, it is not only the child, but the pregnant or nursing mother, who need to be protected from BPA exposure. Since children cannot be responsible for making choices regarding BPA, clear information must be provided to mothers so that they can made the educated choices needed to protect themselves and their children. As we will see, the information is not always clear and not always accessible to the mothers who need it.

Poor Babies and BPA

According to the Centers for Disease Control study, the majority of children and mothers in the U.S. have detectible levels of BPA in their bodies (Calafat et al. 2007). The study also found that there is a disparity of BPA burden associated with socio-economic status. The authors found a strong correlation between lower income and higher BPA in the urine of participants. The correlation was incremental such that participants in the study with household incomes of less than $20,000 had the most BPA, those with household incomes between $20,000 and $45,000 had less BPA, and people with household incomes greater than $45,000 had the least BPA in the urine. Clearly, low socio-economic status is a predictor of high BPA burden. But what is the route of exposure?

A main route of exposure is thought to be through consumption of contaminated food (Thompson et al. 2003; Wilson et al. 2007). A main source of contamination with BPA is the packaging of food in metal cans lined with BPA-containing epoxy resins. The BPA can leach from the lining and into the food (Vandenberg et al. 2007). The Centers for Disease Control BPA-exposure study showed that non-Hispanic blacks carried the highest BPA burden of any ethnicity surveyed (Calafat et al. 2007). African Americans, followed by Hispanics, purchase the most canned produce per capita of any demographic (Buzby et al. 2010). Part of the reason for this may be that in low-income neighborhoods, access to fresh fruits and vegetables at retail outlets is lower. In these areas, food stores are usually small and locally-owned rather than large supermarkets (Hosler 2008). The shelf space allocated to fruits and vegetables (both canned and fresh) at small food stores is significantly less (4 meters) than at a larger supermarket (121 meters). Furthermore, only 32% of that space in small stores is allocated to fresh produce as opposed to 70% in supermarkets (Frazao 2007). Moreover, poor families are more dependent upon food banks, which usually provide more canned goods rather than fresh produce (Frazao 2007; Hosler et al. 2008). As a result, populations in low-income areas have less access to fresh produce and, accordingly, may consume higher levels of canned food, which increases their exposure to BPA. Poor mothers may be exposed to higher levels of BPA and pass this onto their fetuses during pregnancy. They may also be feeding their children more canned food.

This argument, however, does not tell the complete story since Hispanics, as a group, display lower levels of BPA in their urine than non-Hispanic blacks (Calafat et al. 2007). Thus exposure through canned food products does not account for all cases of BPA burden. Indeed, babies from poor households may be exposed the same way as discussed previously—through the use of food containers made with BPA such as bottles and sippy cups. Since BPA leaches at a faster rate if the plastic is old, worn, or scratched, exposure risk may be higher if the containers are not replaced as frequently as in households with higher incomes (Lyons 2000; Brede et al. 2003).

No matter the route of exposure, it is clear that infants and children, particularly those from low-income families, are burdened with the highest levels of BPA. Infants and children are the least effective at clearing BPA from the body. Moreover, as discussed below, they are the most susceptible of all populations to the endocrine disrupting character of BPA.

The Detrimental Health Effects of BPA

Naturally, the prevalence of human exposure to BPA, no matter what the route of exposure may be, leads to questions about safety and negative health effects. So what are the health effects?

BPA is called an endocrine disrupting chemical (EDC) because exposure affects the endocrine, or hormonal, system. Basically, BPA acts like estrogen. In simple terms, estrogen is often thought of as the female sex hormone that causes girls to enter puberty and that is low in women after menopause. In reality, estrogen is a carefully balanced chemical messenger that is found in both men and women. Circulating in very low concentrations, estrogen triggers responses in cells and tissues and is critical during fetal development and in normal health and reproduction. Because BPA can act like estrogen (and, paradoxically, in some cases behave just the opposite of estrogen), exposure to BPA can alter the proper hormonal balance of the body and may lead to detrimental health effects.

Exposure to BPA has been studied on many levels, from effects on individual cells, to tissues and whole organisms. For example, adult male mice and rats exposed to BPA produced abnormal sperm and had reduced fertility; however, these effects were reversed when BPA was no longer administered to the animals (Toyama et al. 2004). Studies in rats indicate that BPA exposure might enhance the risk of developing diabetes and hypertension (Alonso-Magdalena et al. 2006, 2010). While most health studies regarding exposure to BPA are done in laboratory animals, the few human epidemiological studies reveal a relationship between BPA exposure and polycystic ovary syndrome, repeated miscarriage, and heart disease (Takeuchi et al. 2004; Sugiura-Ogasawara 2005; Melzer et al. 2010). Additionally, BPA causes a human breast cancer cell line to proliferate, suggesting that estrogen-sensitive tissues and cells in the human body may react similarly (Singleton 2004; Ho et al. 2006).

Many animal studies focus on the effect of BPA exposure during fetal development as this is a time when cells and tissues are especially susceptible to alterations in the hormonal environment. Not only does BPA disrupt proper functioning of the placenta during gestation, but offspring of animals that are exposed to BPA during pregnancy exhibit many deleterious health effects (Lee et al. 2005). Male offspring have enlarged prostates and malformed urethra (Nikaido et al. 2004; Markey et al. 2005). Moreover, they are at a higher risk of developing prostate cancer. BPA exposure alters the genital tract of female offspring (Nikaido et al. 2004; Markey et al. 2005). Exposed females also enter puberty earlier (Nikaido et al. 2004). Exposure affects brain development and, as a result, some behavioral differences typically seen between males and females are lost in offspring exposed to BPA in the uterus (Rubin et al. 2006). Based on the clear effects of BPA exposure on the development of fetuses in animals, similar effects on human development are plausible. Indeed, BPA has been found in the bloodstream, placenta, cord blood, and fetal blood of humans at levels that are within the range studied in many of the animal models (Schonfelder et al. 2002).

Beyond the immediate developmental effects of BPA exposure in utero, evidence is beginning to elucidate long-term effects of early-life exposure to BPA, potentially through the modification of epigenetic programming via altered DNA methylation (Vom Saal et al. 2007; Kudakovic and Champagne 2011). One study suggests that such detrimental modifications may be counteracted by maternal nutrient supplementation (Dolinoy et al. 2007). Nevertheless, these changes may result in alterations that are passed on transgenerationally, resulting in adverse health effects in subsequent generations who, themselves, may not have been directly exposed to BPA (Kudakovic and Champagne 2011; Walker and Gore 2011).

How Is The "Safe" Limit of BPA Exposure Determined?

Although BPA was not used in the manufacture of plastics until the 1950s, the hormonal activity of BPA was reported in 1936 (Dodds and Lawson 1936). For decades, products containing BPA were shown not to release the compound and thus these products were deemed safe. Indeed, the current United States Environmental Protection Agency (EPA) and Food and Drug Administration (FDA) regulations regarding allowable levels of BPA exposure are based on these early findings.

Let's step back a moment and consider the roles of regulatory agencies such as the FDA and the EPA and how these agencies determine the "safe" human exposure level for a chemical. Founded in 1938, the FDA focuses on ensuring safety of food, drugs, and medical products. Much later, in 1970, the EPA was established to protect human health in general and safeguard the environment by consolidating the varied efforts of research, monitoring, standard-setting, and enforcement. Six years after the creation of the EPA, the Toxic Substances Control Act was passed by Congress. This act gave the EPA the power to control chemicals that pose an unreasonable risk to human health or the environment. In other words, the EPA was charged with determining the "safe" human exposure level for chemicals.

Just like people, in the United States, chemicals are presumed innocent until proven guilty—in other words, safe until proven harmful. Manufacturers of new chemicals are required to provide only a 90-day notice to the EPA prior to production and less than 1% of all new chemicals are required to undergo testing to determine human safety. As a result of this regulatory philosophy, chemicals can be used in manufacturing or commercially that have not been tested for safety. Since taking on the daunting task of monitoring over 80,000 chemicals produced or imported into the United States, the EPA has taken action to reduce the risk of over 3600 chemicals, through such actions as limiting new uses of a compound. Remarkably, the EPA has banned the production or use of only five chemicals (polychlorinated biphenyls (PCBs), halogenated chlorofluoroalkanes, dioxins, hexavalent chromium, and dichloro-diphenyl-trichloroethane (DDT)) (EPA 2001, 2011; Stephenson 2006; Shapiro 2007). So how is the safety of these chemicals determined? As we will see, it is not always a straightforward route and the final results are sometimes ambiguous and misleading.

While certain tests to determine the safety of a chemical can be accomplished in a test tube or culture dish, toxicology studies to determine the health effects of exposure to a chemical require animal studies. Primarily small mammals (such as mice) are used and the results are then extrapolated to humans. The highest level of chemical exposure at which no adverse health effect is seen in the animal tests is the No Observable Adverse Effect Level (NOAEL). Uncertainty remains, however, in assessing the effect of long-term (indeed sometimes life-long or multigenerational) exposure to potentially toxic chemicals. Moreover, since there is variability between species (mice versus humans for example) and variability between people (one person might be particularly sensitive to a chemical) a "factor of ten" approach is used to determine the anticipated "safe dose" for humans. For example, this approach would take the NOAEL from the animal study and divide it by ten to account for species variability and by another ten to account for variability between people and by another ten to account for long-term exposure. Thus the final exposure level deemed safe for humans would be ten times ten times ten, or one thousand times less than that determined in the animal study. This "safe level" is sometimes referred to as the Oral Reference Dose (RfD), the Acceptable Daily Intake (ADI) or the Tolerable Daily Intake (TDI). Currently the EPA lists the RfD for BPA as being 0.05 milligrams (or 0.00005 grams) of BPA per kilogram of body weight per day (EPA 1988). Following this guideline, a person weighing 140 pounds (~63 kg) could "safely" ingest 0.003 grams of BPA per day, or a little over a gram of BPA each year. Accordingly, this level is higher than what a typical adult is exposed to daily, which can range from 0.000008 to 0.0015 milligrams of BPA per kilogram of body weight per day (NTP-CERHR 2008).

More or Less is Better (or Worse)?

The method of determining the NOAEL is grounded in the classical paradigm of toxicology called the threshold dose response. This states that there is a certain dose (the threshold) at which a compound causes a biological effect and that doses lower than the threshold show no effect (Calabrese and Baldwin 2003; Calabrese 2006, 2010). Although this paradigm has generally been regarded to be true, BPA has proven otherwise.

Initially, in the 1980s, low levels of BPA were not even considered as a concern since BPA contamination at very low levels was not seen. However, this was not because the BPA was not present in the sample, for example, in water or food kept in containers made with BPA-containing polycarbonate plastics. Rather, the BPA was there, but the scientific techniques to detect it were not sensitive enough. However, since the late 1990s, scientific techniques progressed such that very small amounts of BPA could finally be measured accurately. Levels as low as parts per billion (ppb) are now routinely detected in the laboratory.

Unfortunately, the ability to simply detect such low levels is often not good enough since tissues and cells can respond to levels of BPA that are 100 times lower (Vom Saal and Welshons 2006). The first such study showing a detrimental effect of BPA at very low doses was published in 1997 and, since then, hundreds have been published (e.g. Vom Saal et al. 1997; Vom Saal and Welshons 2006). In 2000, the Endocrine Disruptors Low Dose Peer Review panel, a 36-member panel formed by the National Institute of Environmental Health Science's National Toxicology Program at the request of the EPA, reviewed the data available at the time and concluded that exposure to astonishingly low doses of BPA causes physiological effects and that further studies on the physiological effects of the compound were warranted (EPA 2000; Kaiser 2000).

The findings of the Endocrine Disruptors Low Dose Peer Review panel go against the threshold dose response paradigm. The majority of studies on the biological effects of BPA show that BPA follows a different dose response model, called the biphasic model, or the "low dose theory". The "low dose theory" asserts that extremely low amounts of BPA exposure result in biological effects that are not seen or are opposite from effects observed at higher doses. As a concrete example of a biphasic model, consider physical exercise. No or very little exercise can result in higher susceptibility to disease, while extreme exercise can also be detrimental. However, a moderate level of exercise is most beneficial. Clearly exercise does not follow the threshold dose response. Accordingly, BPA does not fit that model either since low doses often are more harmful than higher doses. Results confirming the "low dose theory" in regards to BPA are repeatable and accepted by the scientists who specialize in hormonal health effects (Welshons et al. 2006; NTP-CERHR 2008).

Combining the classical NOAEL determination with the "low dose theory" leads to some problems and ambiguity in deciphering the scientific findings. The "low dose theory" states that low doses are often more detrimental than high doses. The NOAEL determines the highest dose that causes no observable adverse effects. Herein lies the problem. The regulatory agencies were designed to determine safety using the threshold dose response for a chemical and the protocols and methodologies are not in place to monitor health effects for chemicals that may not adhere to the classic threshold paradigm. As a result, many chemicals, including BPA, that fit the "low dose theory" were virtually eliminated from consideration by regulatory agencies as a concern for public health because they didn't fit the canonical beliefs regarding the relationship between exposure dose of a chemical and resulting health effects.

The Differences between Basic Research and the Toxicology Testing Industry

At this point, it's necessary to consider two different types of scientific research surrounding BPA—basic research and the toxicology testing industry. Both provide important information about BPA safety and health effects, but the methods and approaches to obtaining the information are different.

The toxicology testing industry follows regimented guidelines set forth by the EPA and/or FDA to evaluate chemical safety. Each new testing procedure used must undergo a period of comment, revision and validation that can take several years (or decades) to complete before it can be put in place. Often the validation process, which involves the generation of a consistent result for an experiment carried out by multiple contract laboratories, is impossible to complete if the assay under investigation is difficult to perform (Borrell 2010). This can result in the elimination of otherwise valid assays that may be critical in determining the safety of a chemical. In addition, industry labs are required to follow quality-control standards that are called Good Laboratory Practices (GLPs). GLPs were first instituted in response to laboratory fraud that was occurring in industry testing laboratories. GLPs standardize data storage and thus ensure that the experiments truly occurred by requiring extensive recordkeeping. GLPs reveal little about the legitimacy of the experimental results because they do nothing to ensure that the results are valid or based on experiments that were done correctly (Vom Saal 2010). The institution of GLPs was very important because the studies carried out by the testing industry are usually not repeated by other scientists or testing facilities. Thus, following GLPs ensures that a proper and complete record of the original study is documented.

The aim of basic research, sometimes referred to as academic science, is not necessarily to test the toxicity of a chemical, but rather to explore the biological mechanisms underlying the activity of a chemical. Such studies do not always follow the guidelines used by the toxicology testing industry and most do not follow GLPs. Maintaining a GLP-compliant laboratory is expensive and time-consuming and not required of basic science because this research is subjected to verification through replication and peer-review (Vom Saal 2010).

The fundamental differences between basic research and the toxicology testing industry have led to even more haziness in the information available about the safety of BPA. While both approaches to the study of BPA are legitimate, the differences have been highlighted by some to debunk valid findings. We will see that this leads to a befuddled slurry of information and misconceptions that have left regulatory agencies, legislatures, the media, and as a result, the common mother, in a state of confusion.

The Confusing BPA Story

Let's review our information at this point about BPA. First, BPA is found in some plastic food containers and in the lining of metal food cans. Small amounts of BPA are released from the containers and ingested by people. Second, small amounts of BPA have been shown to have deleterious effects on health (the "low dose theory"). Third, the current level of BPA exposure considered "safe" by regulatory agencies is much higher than the low doses to which people are routinely exposed.

How is this information regarding BPA and its safety being conveyed to the public? The information is out there, but it is a puzzle to get through. The not-for-profit organization Environmental Health Sciences was founded in 2002 to help increase public understanding of emerging scientific links between environmental exposures and human health (Environmental Health Sciences 2006). In the first four years of monitoring the world press, the organization came across over 100 stories regarding BPA in newspapers and on television. That is not a significant outpouring of information to mothers who need the information. For comparison, over 500 stories regarding tobacco use were published in only the first four months of 2011 (Environmental Health Sciences 2011). The stream of information on BPA has been improving, and over 2000 additional stories on BPA surfaced worldwide in the next four years. As a result, there is now a better chance that a typical mother might actually hear something about BPA in the media. The problem is that many of the stories contain conflicting information on the health effects of BPA and thus their power to influence the behavior of the common person in society is diminished. So what are these conflicts and where are they coming from?

A first conflict involves the level of BPA that is actually released from plastic products, and thus accessible for human uptake. Early studies prior to 1999 indicated that BPA did not leach or leached in very small amounts from plastic products, including baby bottles. These studies are sometimes referred to by those in the chemical industry who have a vested interest in maintaining the use of BPA in plastics production (Polycarbonate/BPA Global Group 2006). For example, the Polycarbonate/BPA Global Group of the American Chemistry Council is the main advocate for the use of BPA and represents plastic manufacturers worldwide. Since 1999 many studies have shown that BPA leaches from products at levels shown to cause health effects in animals. The Polycarbonate/BPA Global Group now consents that some leaching may occur, but maintains that the levels are significantly lower than the RfD set for BPA by the EPA (0.05 milligrams of BPA per kilogram of body weight per day) and thus are no cause for concern (EPA 1988; Polycarbonate/BPA Global Group 2011).

This leads to a second conflict—the doses used to determine any adverse health effects from BPA exposure. Early studies on BPA exposure tended to find little resulting adverse health effects, yet these studies were often designed using doses that were higher than those now regarded as being in an environmentally relevant range—in other words, doses that are closer to the low doses that human are exposed to regularly and that fit the "low dose theory." These were the main studies initially used by the EPA to determine the "safe" level of BPA exposure (the RfD) and that, again, were often referenced by chemical companies to attest to the safety of BPA (EPA 1988). However, over the past two decades, the biphasic dose response of BPA has been proven time and again by basic researchers. Accordingly, the Polycarbonate/BPA Global Group has systematically criticized such research in an attempt to debunk the legitimacy of the findings. Some criticisms were valid, such as the use of a small sample size or analysis of effects at a single dose of BPA. However, the most frequent condemnation pointed to supposed faulty results because GLPs were not followed (Borrell 2010). This is a misleading claim given that GLPs were developed to standardize data storage and reveal little about the legitimacy of the results (Vom Saal 2010).

This leads to a third conflict—how the scientific research is designed and funded. We've already discussed the fundamental difference in design of research carried out by basic researchers and the toxicology testing industry. In 2006 an article in the journal Environmental Research (Vom Saal 2006) presented an argument that there is bias in how BPA studies are designed and that the bias is related to the source funding the study. The authors claim that over 90% of the 109 studies published as of 2005 that were funded by the government (through grants from agencies such as the National Institutes of Health or the National Science Foundation) concluded that observable health effects occur upon exposure to low doses of BPA. Alternatively, they claim a bias in industry-funded research such that 100% of the 11 industry-funded studies concluded that no observable health effects occur. The authors further suggest that the reason for this discrepancy is because the industry-funded research did not follow established standards of experimental design and analysis. Examples include failing to include appropriate controls or using animals with known low sensitivity to estrogens to conclude that no effect is seen. The authors claim that "when the investigators find what those funding the research desire them to find, the fact that in science it is considered a violation of the scientific process to draw positive conclusions from uniformly negative data is simply ignored" (Vom Saal and Welshons 2006).

There is a clear difference between how science is carried out in industry-funded laboratories versus academic laboratories. In most situations, the results of both are correct. Yet, in the case of BPA, battles were taking place. Basic research was mounting more and more data in support of the low-dose theory. The chemical industry, threatened by the possibility of having to reduce or eliminate BPA in some products, continued to try to debunk the basic research. All the while, as we will see, the regulatory agencies were failing to adequately respond to the new findings.

Legislation Tries to Fill in the Regulatory Gap

Clearly, the timing, design, analysis, and funding of the scientific research are all contributors to the ambiguity of the BPA story. Because of this ambiguity, the findings can be obscured by those who disseminate the information to the public, especially those with a vested interest in BPA production and usage. This leads to the presentation of a confusing and unclear picture of the health risks of BPA exposure by the media and others and a resulting lack of comprehension and action by the public and policy makers.

The resulting influence of this ambiguity was revealed in the spring of 2006 when state legislators in California, Maryland and Minnesota attempted to pass legislation that would ban the use of BPA in products aimed at children. None of them passed.

The bills focused on children because their lower body weight causes them to be far more susceptible to adverse effects from chemical exposures than adults, even at very low doses. Moreover, children are more likely to be exposed to BPA orally because of their need to mouth products. Indeed, some products, such as baby bottles and teething rings, which contain BPA, are specifically designed for this purpose.

The California bill (AB319) was introduced in February 2005, making it the first such legislation to be introduced in any state. Sponsored by Assembly Member Wilma Chan (Democrat representing Alameda), AB319 called for any products, including toys or childcare articles, intended for use by a child 3 years of age or younger that contain BPA to be prohibited in the state. These articles would not be allowed to be manufactured, sold, or distributed in any way. The bill also called for the ban of child-aimed products that contain other harmful chemicals such as phthalates. Civil action against violators of the ban would be carried out by the Attorney General and could result in penalties of no less than $10,000 for each day of violation (California Assembly Bill No. 319 2005). The Fact Sheet accompanying the bill put it in plain words,

"AB319 recognizes that we must act now to prevent exposure by eliminating at the source the chemicals, such as Bisphenol-A and Phthalates that pollute our bodies. By making intelligent decisions about what chemicals we allow into the environment, we can prevent unnecessary exposures to dangerous substances. Furthermore, children are incredibly sensitive to chemical pollution. Their developing organs and systems, as well as their natural behavior, put them at tremendous risk. Some chemicals are simply too toxic and dangerous to children, to allow exposures to continue."

Given the broad economic ramifications of the bill, it was energetically opposed by stakeholders in the chemical, plastics, baby products, and grocery industries. Under the umbrella organization Coalition for Consumer Choice, the NoAB319 campaign successfully fought the bill both in the media and in the assembly hearing. In a news release by NoAB319, Steve Hentges, now executive director of the Polycarbonate/BPA Global Group of the American Chemistry Council, stated that the legislation was "founded on insubstantial claims and unproven hypotheses that lack scientific rigor." This assertion was apparently focused on the "low-dose theory" of BPA exposure. Such "doubt" tactics are reminiscent of the tobacco industry's campaign to debunk the scientific findings regarding the harm caused by cigarette smoke. As one cigarette executive wrote in a confidential industry memo "Doubt is our product since it is the best means of competing with the 'body of fact' that exists in the minds of the general public. It is also the means of establishing a controversy" (Brown and Williamson 1969).

Clearly, the contradictory information set forth by the proponents and opponents of AB319 yielded concern. Indeed, the bill died in the Appropriations Committee, even after an amendment removed the BPA provisions, because of one deciding vote. San Francisco Democrat Leland Yee, according to a spokesman, "decided that the decisions to ban chemicals should be left to health experts, not politicians, especially after scientists gave conflicting testimony at an Assembly hearing last week" (Cone 2006).

While this same story has repeated itself in legislation put forth in other states, some legislation limiting BPA in products aimed at children has been successful. As of 2010, eight states have banned BPA in baby bottles—Connecticut, Maryland, Massachusetts, Minnesota, New York, Vermont, Washington, and Wisconsin.

At a more local level, the first legislation to ban BPA from products aimed at children passed in the city of San Francisco. The "Stop Toxic Toys" bill was virtually identical to AB319 and was signed into law on June 16, 2006 (Goodyear 2006). However, before the bill could be enforced, it was repealed the following year. Since then, some other municipalities, including Chicago, have successfully passed legislation banning BPA from baby bottles.

While the local and state legislation are important steps, a piece-meal approach to controlling BPA exposure, especially in young children, is not perfect. Companies and businesses are bound to have difficulty conforming to a variety of regulations. Although BPA-free alternatives for products aimed at infants and young children are often available, consumers in areas with legislation may find a lack of choices when it comes to plastic products on the store shelves. Ideally the national regulatory agencies should step in to minimize these problems.

Taking a Second Look at the Science

Given the debate surrounding the science behind BPA, an expert panel was convened in Chapel Hill, NC, in 2006 to carry out an "examination of the relevance of ecological, in vitro and laboratory animal studies for assessing risks to human health." The panel meeting, sponsored by the National Institutes of Health (NIH), concluded that "the wide range of adverse effects of low doses of BPA in laboratory animals exposed both during development and in adulthood is a great cause for concern with regard to the potential for similar adverse effects in humans" (Vom Saal et al. 2007).

Shortly after, in 2008, the NIH, working through the National Toxicology Program (NTP), convened another panel of scientists to carry out a comprehensive review of the scientific data on BPA exposure (both in animals and humans and carried out by both industrial and academic laboratories). The Center for the Evaluation of Risks to Human Reproduction (CERHR) panel published a monograph that focused on the potential human reproductive and developmental effects of BPA exposure (NTP-CERHR 2008). The NTP/CERHR panel applied a five-level scale of concern ("negligible," "minimal," "some," "concern," and "serious concern") in order to rate the exposure risk of BPA for various populations. As a point of reference, the mid-point in the scale, "some" concern, is the same level that the NTP earlier applied to the exposure of fetuses to amphetamines (Wiles 2010). The panel expressed "some" or "minimal" concern that fetal exposure to BPA causes neural, behavioral, and prostate problems, as well as accelerations in puberty. They stated "negligible" concern that BPA would cause birth defects or malformations and, when considering infants and children, "some" concern that neural and behavioral effects may occur and "minimal" concern for accelerations in puberty. For adult populations, only those subgroups exposed to higher levels of BPA, for example through occupational exposure, were highlighted as having minimal risk for adverse reproductive effects. Overall, the panel concluded that "the possibility that human development may be altered by bisphenol A at current exposure levels cannot be dismissed" (NTP-CERHR 2008).

Following the NTP/CERHR report, Canada was the first country to take a dramatic step toward limiting BPA exposure in children. In 2008 Canada banned the use of polycarbonate baby bottles (Canada Gazette 2010). Denmark has followed the lead and, in 2010, enacted a temporary ban on BPA in baby bottles, sippy cups, and packaging containers for baby food and formula (EWG 2010). Where does the rest of Europe stand on the issue? In 2002 the European Food Safety Authority (EFSA), after considering the then-available science on BPA, temporarily lowered the "safe" level of exposure to BPA to 0.01 milligrams per kilogram of body weight per day. This was five times lower than the EPA "safe" level (RfD). In 2006, however, the EFSA raised the "safe" level back up to 0.05 milligrams per kilogram of body weight per day after considering the scientific evidence available at the time, that the exposure level of all populations is less than the proposed RfD, and that a temporary RfD was no longer appropriate (EFSA 2006, 2007). Even after considering the U.S. NTP/CERHR draft brief on BPA in 2008, the EFSA held the RfD at the higher level (EFSA 2008).

United States Regulatory Agencies Admit Concern Regarding BPA

Humans are typically exposed to about 0.001 milligrams of BPA per kilogram of body weight per day. This is 50 times lower than the EPA-deemed and FDA-deemed "safe" limit. Unfortunately, this level of exposure is still significantly higher than the low doses that have been shown to cause adverse health effects. The concern is highest when fetuses, infants, and children are considered. Given these concerns and the lack of U.S. regulatory response to the published results of the NTP/CERHR, the U.S. House of Representatives Committee on Energy and Commerce urged the FDA to reconsider the safety of BPA (CEC 2009). After initially ignoring the NTP/CERHR findings, the FDA is now reevaluating its risk assessments of BPA that takes into consideration infant exposures at very low doses. This new approach was initiated after the FDA process was found to be flawed after review by an FDA BPA subcommittee (FDA SBSOBA 2008). For example, the subcommittee determined that some valid BPA studies had not been included and that the level of BPA exposure of infants was underestimated. Ultimately the subcommittee concluded that there is "scientific support for the use of a point of departure substantially below (i.e., at least one or more orders of magnitude lower than) the 0.05 mg/kg bw/day level selected in the draft FDA assessment" (FDA SBSOBA 2008). In other words, the subpanel concluded that the current RfD is too high.

In 2010, for the first time, a U.S. regulatory agency, the FDA, admitted some concern over the safety of BPA and is now investigating the uncertainties surrounding BPA. The EPA also added BPA to its "Chemicals of Concern" list under the Toxic Substances Control Act (EPA 2010). Ideally these moves will ultimately lead to a reduction in the RfD. The ramifications of such a change could prompt the plastics industry and manufacturers of products containing BPA to reevaluate the use of BPA in their products. Luckily, some manufacturers have already begun removing BPA from their products and consumers can identify such products by their "BPA-free" label. Moreover, a few large retailers, including Toys-R-Us and Wal-Mart, no longer keep BPA-containing baby bottles on their shelves. But is this enough to protect the most vulnerable of our population?

Has BPA Forged a New Way of Doing Science?

BPA is still a controversial chemical. It clearly has valid and important uses in our society. No one will argue that its use in bullet proof shields and protective eyewear is not vital and should be discontinued. Yet, exposure to very small amounts of BPA is harmful, especially to the youngest and most vulnerable of our society. The regulatory system needs to confidently review all research, whether obtained by basic researchers or the toxicology testing industry. Yet, these methods of research vary in their goals and approach. Have we learned anything from this BPA story?

The National Institute of Environmental Health Sciences (NIEHS) is doing its part. They committed thirty million dollars for 2010 and 2011 toward basic research on BPA. But unlike in the past, research projects were chosen that would specifically fill gaps of knowledge on the safety and health effects of BPA. Moreover, the funded scientists were brought together at a single meeting to come to a consensus on how to design and carry out the experiments. In essence, this group of forty scientists was designing the first "BPA master experiment." They would use consistent doses of BPA, share samples, and monitor additional variables (Borrell 2010). This is the first time that such a master experiment of basic research has taken place. Jerry Heindel, the BPA program manager at the NIEHS, said "Let's learn from the lessons of BPA and start developing collaborations and interactions to move the field right from the start" (Borrell 2010).

Although the final word on the safety of BPA has still not been decided, perhaps the lessons have been learned. The "master experiment" approach is now underway for the initial safety studies of nanomaterials (Borrell 2010).

The Information Available is Still Full of Confusion

In the meantime, while we wait for the national regulatory agencies to finally move on limiting BPA exposure, we still need information to make our own decisions. In 2010, the FDA parent agency, the United States Department of Health and Human Services posted on its website some useful tips for parents on how to limit BPA exposure, and some media outlets have published similar lists (USDHHS 2010). While these lists are referred to on the website of the Polycarbonate/BPA Group of the American Chemistry Council, they are presented in the context that the FDA does not recommend avoiding the use of polycarbonate products in food containers. This presentation of mixed messages—it is "safe" but here's how to avoid exposure—makes it difficult for the common person to decide a course of action regarding BPA products.

Such mixed messages may become more prevalent as the BPA stakeholders counteract the flow of information to the public that BPA is unsafe. Indeed, lobbyists for the chemical companies that make and use BPA and food-packaging executives were reported to have met in 2009 to come up with strategies to counteract the current information available to the public that BPA is harmful (Rust and Kissinger 2009). One potential approach included finding a pregnant woman to act as a spokesperson for the beneficial uses of BPA. Such a spokesperson would appeal to the population most in need of understanding the risks associated with BPA exposure of infants and children—the pregnant women and mothers. A second strategy was to generate media stories, aimed primarily at Hispanics and African Americans in poor communities that would claim that canned goods not lined with BPA would more likely be contaminated with bacteria. This claim is false as several BPA-free alternatives for can linings exist that provide protection against contamination (EWG 2007). What is the current risk of exposure to BPA to low-income populations and why are the chemical and food-packaging industries focusing on canned goods? We've already discussed how the Centers for Disease Control assessment of BPA exposure in different populations found that children were the most burdened with contamination (Calafat et al. 2007). It is clear that not only are children a vulnerable population with regards to BPA, but that low-income populations are as well. Mothers and people of low economic status need access to valid information in order to make educated decisions. The information out there is confusing and often misleading. A direct, educational outreach campaign aimed at parents to clarify the facts and provide useful information about lifestyle and consumer choices may be one way to reduce the burden that the infants and children are carrying. Health practitioners, especially obstetricians and pediatricians, also need to be educated about BPA so that they can discuss lifestyle choices with their patients. Ultimately, a course of action to protect all populations would be for the regulatory agencies to lower the "safe" level of BPA exposure. This would require the reduction of the use of BPA not only in products aimed at children, but also in all food packaging.

So, as the mother-in-me still waits anxiously for the regulatory agencies and the legislature to catch up with the research on BPA that the scientist-in-me understands, I still use baby bottles that don't contain BPA and I rarely serve canned vegetables to my son. Nevertheless, while I feel proactive as I watch my son happily drink from his BPA-free bottle, I still cringe a little bit when he drops the baba, toddles over to his toy bin and starts to gnaw on his plastic train instead!

Acknowledgements

The authors would like to thank Aimee Quitmeyer for her assistance in the original information gathering for this essay. Rebecca Roberts would like to thank her sister Catherine and her husband James for support and conversations that helped in developing this chapter.

References

Alonso-Magdalena, P., S. Morimoto, C. Ripoll, E. Fuentes, and A. Nadal, 2006. The estrogenic effect of bisphenol A disrupts pancreatic beta-cell function in vivo and induces insulin resistance. Environmental Health Perspectives, v. 114, pp. 106-112.

Alonso-Magdalena, P., E. Vieira, S. Soriano, L. Menes, D. Burks, I. Quesada, and A. Nadel, 2010 Bisphenol A exposure during pregnancy disrupts glucose homeostasis in mothers and adult male offspring. Environmental Health Perspectives, v. 118, pp. 1243-1250.

Biedermann, S., P. Schudin, and K. Grob, 2010. Transfer of bisphenol A from thermal printer paper to the skin. Analytical and Bioanalytical Chemistry, v. 398, pp. 571-576.

Borrell, B., 2010. Toxicology: The big test for bisphenol A. Nature, v. 464, pp.1122-1124.

Boyd, G. R., J. M. Palmeri, S. Zhang, and D. A. Grimm, 2004. Pharmaceuticals and personal care products (PPCPs) and endocrine disrupting chemicals (EDCs) in storm water canals and Bayou St. John in New Orleans, Louisiana, USA. Science of the Total Environment, v. 333, pp. 137-148.

Brede, C., P. Fjeldal, I. Skjevrak, and H. Herikstad, 2003. Increased migration levels of bisphenol A from polycarbonate baby bottles after dishwashing, boiling and brushing. Food Additives and Contaminants, v. 20, pp. 684-689.

Brown and Williamson, 1969. Smoking and Health Proposal. Brown and Williamson Document No. 680561778-1786. Available online at http://legacy.library.ucsf.edu/tid/nvs40f00.

Buzby, J. C., H. F. Wells, A. Kumcu, B-H. Lin, G. Lucier, and A. Perez, 2010. Canned Fruit and Vegetable Consumption in the United States: An Updated Report to Congress. U. S. Department of Agriculture, Economic Report Service. Available online at http://www.ers.usda.gov/Publications/AP/AP050/AP050.pdf.

Calabrese, E. J. and L. A. Baldwin, 2003. The hermetic dose-response model is more common than the threshold model in toxicology. Toxicological Sciences, v. 71, pp. 246-250.

Calabrese, E. J., 2006. The failure of dose-response models to predict low dose effects: A major challenge for biomedical, toxicological and aging research. Biogerontology, v. 7, pp. 119-122.

Calabrese, E. J., 2010. Hormesis is central to toxicology, pharmacology and risk assessment. Human and Experimental Toxicology, v. 29, p. 249-261.

Calafat, A. M., Z. Kuklenyik, J. A. Reidy, S. P. Caudill, J. Ekong, and L. L. Needham, 2005. Urinary concentrations of bisphenol A and 4-nonylphenol in a human reference population. Environmental Health Perspectives, v. 113, pp. 391-395.

Calafat, A. M., X. Ye, L.-Y. Wong, J. A. Reidy, L. L. Needham, 2007. Exposure of the U.S. population to Bisphenol A and 4-tertiary-Octylphenol: 2003-2004. Environmental Health Perspectives, v. 116, pp. 39-44.

Calafat A. M., J. Weuve, X. Ye, L. T. Jia, H. Hu, S. Ringer, K. Huttner, and R. Hauser, 2009. Exposure to bisphenol A and other phenols in neonatal intensive care unit premature infants. Environmental Health Perspectives, v. 117, pp. 639-644.

Canada Gazette, 2010. Order Amending Schedule I to the Hazardous Products Act (Bisphenol A). Government of Canada. Available online at http://www.gazette.gc.ca/rp-pr/p2/2010/2010-03-31/html/sor-dors53-eng.html.

CEC (Committee on Energy and Commerce), 2009. Energy and Commerce Chairmen Waxman and Stupak Request FDA Review of Bisphenol A (BPA) Decision. Available online at http://democrats.energycommerce.house.gov/index.php?q=news/energy-and-commerce-chairmen-waxman-and-stupak-request-fda-review-of-bisphenol-a-bpa-decision.

Cone, M., 2006. Ban on use of toxic materials in baby products founders. Los Angeles Times, Los Angeles, CA.

Coors, A., P. D. Jones, J. P. Giesy, and H. T. Ratte, 2003. Removal of estrogenic activity from municipal waste landfill leachate assessed with a bioassay based on reporter gene expression. Environmental Science and Technology, v. 37, pp. 3430-3434.

Dodds, E. C. and W. Lawson, 1936. Synthetic oestrogenic agents without the phenanthrene nucleus. Nature, v. 137, p. 996.

Dolinoy, D. C., D. Huang, and R. L. Jirtle, 2007. Maternal nutrient supplementation counteracts bisphenol A-induced DNA hypomethylation in early development. Proceedings of the National Academy of Sciences, v. 104, pp. 13,056-13,061.

EFSA (European Food Safety Authority), 2006. Opinion of the scientific panel on food additives, flavourings, processing aids and materials in contact with food on a request from the commission related to 2,2-bis(4-hydroxyphenyl)propane (Bisphenol A), Question number EFSA-Q-2005-100. The EFSA Journal, v. 428, pp. 1-6.

EFSA (European Food Safety Authority), 2007. EFSA re-evaluates safety of bisphenol A and sets Tolerable Daily Intake. Available online at http://www.efsa.europa.eu/en/press/news/afc070129.htm.

EFSA (European Food Safety Authority), 2008. EFSA updates advice on bisphenol. Available online at http://www.efsa.europa.eu/en/press/news/cef080723.htm.

Environmental Health Sciences, 2006. Environmental Health News. Available online at http://www.environmentalhealthnews.org/.

Environmental Health Sciences, 2011. Environmental Health News. Available online at http://www.environmentalhealthnews.org/.

EWG (Environmental Working Group), 2007. Bisphenol A: Toxic Plastics Chemical in Canned Food: Companies Reduced BPA Exposures in Japan. Available online at http://www.ewg.org/node/20938.

EWG (Environmental Working Group), 2009. Pollution in People: Cord Blood Contaminants in Minority Newborns. Available online at http://www.ewg.org/files/2009-Minority-Cord-Blood-Report.pdf.

EWG (Environmental Working Group), 2010. Timeline: BPA from Invention to Phase-Out. Available online at http://www.ewg.org/reports/bpatimeline.

EPA (U.S. Environmental Protection Agency), 1988. Integrated Risk Information System: Bisphenol A (CASRN 80-05-7). Available online at http://www.epa.gov/iris/subst/0356.htm.

EPA (U.S. Environmental Protection Agency), 2000. Low Dose Endocrine Disruptors Peer Review. National Institute of Environmental Health Sciences, N.N.T.P. Sheraton Imperial Hotel and Convention Center, Research Triangle Park, NC.

EPA (U. S. Environmental Protection Agency), 2001. Arsenic rule benefits analysis: An SAB review. U. S. Environmental Protection Agency Science Advisory Environmental Board. Available online at: http://www.gpoaccess.gov/harvesting/arsenic.pdf.

EPA (U.S. Environmental Protection Agency), 2010. Bisphenol A (BPA) action plan summary. Available online at: http://www.epa.gov/oppt/existingchemicals/pubs/actionplans/bpa.html.

EPA (U.S. Environmental Protection Agency), 2011. Available online at http://www.epa.gov/.

FDA SBSOBA (Food and Drug Administration Science Board Subcommittee on Bisphenol A), 2008. Scientific Peer-Review of the Draft Assessment of Bisphenol A for use in Food Contact Applications. Available online at http://www.fda.gov/ohrms/dockets/ac/08/briefing/2008-4386b1-05.pdf.

Frazao, E., 2007. Food Spending Patterns of Low-Income Households: Will Increasing Purchasing Power Result in Healthier Food Choices? Economic Information Bulletin Number 29-4, Economic Report Services, U. S. Department of Agriculture, Washington, D.C.

Goodson, A., H. Robin, W. Summerfield, and I. Cooper, 2004. Migration of bisphenol A from can coatings - effects of damage, storage conditions and heating. Food Additives and Contaminants, v. 21, pp. 1015-1026.

Goodyear, C., 2006. Board bans chemicals that may harm infants. San Francisco Chronicle, San Francisco, CA.

Green R., R. Hauser, A. M. Calafat, J. Weuve, T. Schettler, S. Ringer, K. Huttner, and H. Hu, 2005. Use of di(2-ethylhexyl) phthalate-containing medical products and urinary levels of mono(2-ethylhexyl)phthalate in neonatal intensive care unit infants. Environmental Health Perspectives, v. 113, pp.1222-1225.

Ho, S. M., W.Y. Tang, J. Belmonte de Frausto, and G. S. Prins, 2006. Developmental exposure to estradiol and bisphenol A increases susceptibility to prostate carcinogenesis and epigenetically regulates phosphodiesterase type 4 cariant 4. Cancer Research, v. 66, pp. 5624-5632.

Hosler, A. S., D. T. Rajulu, B. L. Fredrick, and A. E. Ronsani, 2008. Assessing retail fruit and vegetable availability in urban and rural underserved communities. Preventing Chronic Disease, v. 5. Available online at http://www.cdc.gov/pcd/issues/2008/oct/07_0169.htm.

Houlihan, J., 2007. Toxic Plastics Chemical in Infant Formula. Environmental Working Group. Available online at http://www.ewg.org/reports/bpaformula.

Jackson, J. and R. Sutton, 2008. Sources of endocrine-disrupting chemicals in urban wastewater, Oakland, CA. Science of the Total Environment, v. 405, pp. 153-160.

Kaiser, J., 2000. Endocrine disrupters. Panel cautiously confirms low-dose effects. Science, v. 290, pp. 695-697.

Kolpin, D. W., E. T. Furlong, M. T. Meyer, E. M. Thurman, S. D. Zaugg, L. B. Barber, and H. T. Buxton, 2002. Pharmaceuticals, hormones, and other organic wastewater contaminants in U.S. streams. Environmental Science and Technology, v. 36, pp. 1202-1211.

Kuch, H. M. and K. Ballschmiter, 2001. Determination of endocrine-disrupting phenolic compounds and estrogens in surface and drinking water by HRGC-(NCI)-MS in the picogram per liter range. Environmental Science and Technology, v. 36, pp. 3201-3206.

Kudakovic, M. and F. A. Champagne, 2011. Epigenetic perspective on the developmental effects of bisphenol A. Brain, Behavior, and Immunity, v. 25, pp. 1084-1093. Available online at doi: 10.1016/j.bbi.2011.02.005.

Kuo, H. W. and W. H. Ding, 2004. Trace determination of bisphenol A and phytoestrogens in infant formula powders by gas chromatography-mass spectrometry. Journal of Chromatography A, v. 1027, pp. 67-74.

Kuruto-Niwa, R., Y. Tateoka, Y. Usuki, and R. Nozawa, 2007. Measurement of bisphenol A concentrations in human colostrum. Chemosphere, v. 66, pp. 1160-1164.

Lee, C. K., S. H. Kim, D. H. Moon, J. H. Kim, B. C. Son, D. H. Kim, C. H. Lee, H. D. Kim, J. W. Kim, J. E. Kim, and C. U. Lee, 2005. Effects of bisphenol A on the placental function and reproduction in rats. Journal of Preventive Medicine and Public Health, v. 38, pp. 330-336.

Loos, R., G. Hanke, and S. J. Eisenreich, 2003. Multi-component analysis of polar water pollutants using sequential solid-phase extraction followed by LC-ESI-MS. Journal of Environmental Monitoring, v. 5, pp. 384-394.

Lyons, G., 2000. Bisphenol A: A Known Endocrine Disruptor. WWF European Toxics Programme, Godalming, Surrey, UK.

Markey, C. M., P. R. Wadia, B. S. Rubin, C. Sonnenscheine, and A. M. Soto, 2005. Long term effects of fetal exposure to low doses of the xenoestrogen Bisphenol-A in the female mouse genital tract. Biology of Reproduction, v. 72, pp. 1344-1351.

Melzer D., N. E. Rice, C. Lewis, W. E. Henley, and T. S. Galloway, 2010. Association of urinary Bisphenol A concentration with heart disease: Evidence from NHANES 2003/06. PLoS ONE, v. 5. Available online at http://www.plosone.org/article/info:doi/10.1371/journal.pone.0008673.

Mendum, T., E. Stoler, H. VanBenschoten, and J. C. Warner, 2011. Concentration of bisphenol A in thermal paper. Green Chemistry Letters and Reviews, v. 4, pp. 81-86. Available online at doi: 10.1080/17518253.2010.502908.

Munguia-Lopez, E. M., S. Gerardo-Lugo, E. Peralta, S. Bolumen, and H. Soto Valdez, 2005. Migration of bisphenol A (BPA) from can coatings into a fatty-food stimulant and tuna fish. Food Additives and Contaminants, v. 22, pp. 892-898.

Nikaido, Y., K. Yoshizawa, N. Danbara, M. Tsujita-Kyutoku, T. Uri, N. Uehara, and A. Tsubura, 2004. Effects of maternal xenoestrogen exposure on development of the reproductive tract and mammary gland in female CD-1 mouse offspring. Reproductive Toxicology, v. 18, pp. 803-811.

NTP-CERHR (National Toxicology Program-Center for the Evaluation of Risks to Human Reproduction), 2008. NTP-CERHR Monograph on the Potential Human Reproductive and

Developmental Effects of Bisphenol A. NIH Publication No. 08-5994. U.S. Department of Health and Human Services, National Institutes of Health, National Toxicology Program, Center for the Evaluation of Risks to Human Reproduction. Available online at http://oehha.ca.gov/prop65/CRNR_notices/state_listing/data_callin/pdf/NTP_CERHR_0908_bisphenolA.pdf.

Polycarbonate/BPA Global Group, 2006. Bisphenol A. Available online at http://www.bisphenol-a.org.

Polycarbonate/BPA Global Group, 2011. Bisphenol A. Available online at http://factsaboutbpa.org/.

Rubin, B. S., J. R. Lenkowski, C. M. Schaeberle, L. N. Vandenberg, P. M. Ronsheim, and M. Soto, 2006. Evidence of altered brain sexual differentiation in mice exposed perinatally to low, environmentally relevant levels of bisphenol A. Endocrinology, v. 147, pp. 3681-3691.

Rust, S. and M. Kissinger, 2009. BPA industry seeks to polish image. Journal Sentinel, Milwaukee, WI.

Schonfelder, G., W. Wittfoht, H. Hopp, C. E. Talsness, M. Paul, and I. Chahoud, 2002. Parent bisphenol A accumulation in the human maternal-fetal-placental unit. Environmental Health Perspectives, v. 110, pp. A703-A707.

Shapiro, M., 2007. Exposed: The Toxic Chemistry of Everyday Products and What's at Stake for American Power. Chelsea Green Publishing, White River Junction, VT.

Singleton, D.W., Y. Feng, Y. Chen, S. J. Busch, A. V. Lee, A. Puga, and S. A. Khan, 2004. Bisphenol-A and estradiol exert novel gene regulation in human MCF-7 derived breast cancer cells. Molecular and Cellular Endocrinology, v. 221, pp. 47-55.

Stephenson, J. B., 2006. Chemical Regulation, Actions are Needed to Improve the Effectiveness of EPA's Chemical Review Program. Testimony before the Committee on Environment and Public Works, U.S. Senate. United States Government Accountability Office.

Sugiura-Ogasawara, M., Y. Ozaki, S. Sonta, T. Makino, and K. Suzumori, 2005. Exposure to bisphenol A is associated with recurrent miscarriage. Human Reproduction, v. 20, pp. 2325-2329.

Sun, Y., M. Irie, N. Kishikawa, M. Wada, N. Kuroda, and K. Nakashima, 2004. Determination of bisphenol A in human breast milk by HPLC with column-switching and fluorescence detection. Biomedical Chromatography, v. 18, pp. 501-507.

Takeuchi, T., O. Tsutsumi, Y. Ikezuki, Y. Takai, and Y. Taketani, 2004. Positive relationship between androgen and the endocrine disruptor, bisphenol A, in normal women and women with ovarian dysfunction. Endocrinology Journal, v. 51, pp. 165-169.

Thompson, B. M., P. J. Cressey, and I. C. Shaw, 2003. Dietary exposure to xenoestrogens in New Zealand. Journal of Environmental Monitoring, v. 5, pp. 229-235.

Toyama, Y., F. Suzuki-Toyota, M. Maekawa, C. Ito, and K. Toshimori, 2004. Adverse effects of bisphenol A to spermiogenesis in mice and rats. Archives of Histology and Cytology, v. 67, pp. 373-381.

USDHHS (U.S. Department of Health and Human Services), 2010. Bisphenol A (BPA) Information for Parents. Available online at http://www.hhs.gov/safety/bpa/.

Vandenberg, L. N., R. Hauser, M. Marcus, N. Olea, and W. V. Welshons, 2007. Human exposure to bisphenol A (BPA). Reproductive Toxicology, v. 24, pp. 139-177.

Vom Saal, F. S., B. G. Timms, M. M. Montano, P. Palanza, K. A. Thayer, S. C. Nagel, M. D. Dhar, V. K. Ganjam, S. Parmigiani, and W. V Welshons, 1997. Prostate enlargement in mice due to fetal exposure to low doses of estradiol or diethylstilbestrol and opposite effects at high doses. Proceedings of the National Academy of Science, USA, v. 94, pp. 2056-2061.

Vom Saal, F. S., C. A. Richter, R. R. Ruhlen, S. C. Nagel, B. G. Timms, and W. V. Welshons, 2005. The importance of appropriate controls, animal feed, and animal models in interpreting results from low-dose studies of bisphenol A. Birth Defects Research A. Clinical and Molecular Teratology, v. 73, pp. 140-145.

Vom Saal, F. S. and W. V. Welshons, 2006. Large effects from small exposures. II. The importance of positive controls in low-dose research on bisphenol A. Environmental Research, v. 100, pp. 50-76.

Vom Saal, F. S. et al., 2007. Chapel Hill bisphenol A expert panel consensus statement: Integration of mechanisms, effects in animals and potential to impact human health at current levels of exposure. Reproductive Toxicology, v 24, pp. 131-138.

Vom Saal, F., 2010. Presentation on House Bill 221. Testimony to the Commonwealth of Pennsylvania House of Representatives Consumer Affairs Committee Hearing, pp. 96-111.

Walker, D. M. and A. C. Gore, 2011. Transgenerational neuroendocrine disruption of reproduction. National Review of Endocrinology, v. 7, pp. 197-207.

Welshons, W. V., S. C. Nagel, and F. S. Vom Saal, 2006. Large effects from small exposures. III. Endocrine mechanisms mediating effects of bisphenol A at levels of human exposure. Endocrinology, v. 147, pp. S56-S69.

Wiles, R., 2010. Presentation on House Bill 221. Testimony to the Commonwealth of Pennsylvania House of Representatives Consumer Affairs Committee Hearing, p. 47.

Wilson, N. K., J. C. Chuang, M. K. Morgan, R. A. Lordo, and L. S. Sheldon, 2007. An observational study of the potential exposures of preschool children to pentachlorophenol, bisphenol-A, and nonylphenol at home and daycare. Environmental Research, v. 103, pp. 9-20.

Chapter 4: "We Need More than Thirty Trees": The Case for Urban Forestry
Sarah A. Levy, Paula Randler, and Dana Coelho

Summary

Urban forestry involves more than simply planting street trees; it requires an understanding of the complex relationships among wildlife and flora, humans and habitat, and politics and power. In West Oakland, CA, for example, trees are not just part of an urban ecosystem that includes shrubs, grass, gardens, birds, insects, and people. Nor are they are not simply biological engines that produce oxygen, clean pollutants out of the air we breathe, filter stormwater, and provide habitat for diverse species. They are symbols of community, safety, and progress. Planting trees and caring for the urban forest is a way to rally the poverty-stricken around a positive, healthy future. These activities can help lower high asthma rates, provide jobs for local youth, reduce crime, and increase energy savings. In West Oakland, as in other communities around the United States, trees are not an end in themselves. They provide the means to save lives.

Introduction

"See that corner over there?" Kemba Shakur, founder and director of Urban Releaf, pointed to the rough edge of a sidewalk corner connecting a main thoroughfare and smaller residential street. "A kid just got shot there last month."

Kemba shook her head. "He'd just applied to work for us too—had submitted his application only a couple of months ago. And now he's dead."

I stared at the street out the window of one of Urban Releaf's big dark work vans. The sidewalk seemed normal enough, dirty-white in color, exceedingly un-menacing and sidewalk-ish. We'd been driving through West Oakland, CA, for the past half hour, Kemba my unflappable tour guide, showing me what life was like in one of the poorest, most violent places in the Bay Area. Kemba was professional super-mom and neighborhood crusader and seemed equally comfortable writing lengthy grant applications and waging her own version of environmental guerilla warfare against the poverty, crime, and high unemployment rates in her own zip code. Almost a decade ago, Kemba had started Urban Releaf to plant trees in a neighborhood that had, at one time, been almost completely devoid of foliage. She seemed uniquely suited to the job. With long dreadlocks, a quick smile, and intimate knowledge of the language and culture of the West Oakland streets, Kemba could command respect from residents without posing a threat to the various factions in ever-present turf wars. Kemba also made sure I knew that her organization was one of the only urban forestry groups in the nation that was founded and run by a Black woman, specifically devoted to hiring and planting trees in a minority community.

Much of the tour through the neighborhood consisted of Kemba pointing out shuttered houses, sidewalk corners with grisly histories, and—always the highlight—street trees planted by her organization.

"See these over here?" Kemba would nod excitedly towards a row of young saplings struggling towards the sky from a patch of soil in the sidewalk. "They're ours. They belong to Urban Releaf."

Towards the end of one block, Kemba slowed her van down to a crawl, pointing out empty patches of soil that looked like they had once held her trees. Kemba told me that sometimes after her organization planted trees, a few weeks later the trees would be gone, mostly likely succumbing as scapegoats to the simmering rage of the community.

"I think some people like to take their anger out on trees," Kemba said with a shrug. "If people have a difficult time surviving here, why should the trees be able to?"

A few yards from where the trees had been removed stood three teenage boys huddled together, their body language implying that they were at once disinterested and hyper-aware of the dark van inching towards them. Hands in pockets, heads down, eyes up, their bodies forming a horseshoe on the sidewalk. As we rolled up next to them, all three raised their heads to look at us.

Kemba stopped the van and stared at them, and seemed to search their faces for a hint of recognition. They stared back. After a few seconds of mutual unabashed voyeurism, Kemba rolled down my window and leaned across the van to yell out.

"Hey!" She called to the boys. They looked at each other. "Hey! C'mere!"

They looked at each other skeptically. Kemba called out to them again. "Hey! Yeah! I'm talking to you! Come here!"

One of the boys walked over, attempting to look decidedly disinterested. He sauntered up to the car. "Yeah?"

"Hey, do you know what happened to our trees?" Kemba asked pointedly.

The kid shrugged.

"Hey now, I know you know!" Kemba gestured to the empty patches of soil. "I planted three trees on this street and they're all gone. You know what happened to our trees?"

He shrugged again, then shook his head.

One of the other boys approached the van and peered inside. "Hey," he said to Kemba. "You're that tree lady, right?"

Kemba nodded. "Yeah, I'm the tree lady. My organization's named Urban Releaf, and we plant trees in this neighborhood. And we're gonna come back to this street and replant the trees that have been cut down. And I don't want anything to happen to them, you understand? You've gotta watch them for me."

The second boy nodded. "Yeah, I've seen you around. You guys are always planting trees."

"You gonna watch the neighborhood trees for me?" Kemba asked.

The two kids looked at each other and shrugged, and gave a little wave and walked away.

"I'll be back! Watch our trees!" Kemba yelled after them. She rolled up the window and moved the car forward. She shook her head. "It's important for these kids to take care of the trees on their own streets. We can't invest in the trees without investing in the community too. The chance of these trees surviving depends on community support."

She shook her head again. "What you've gotta understand, Sarah, is that we're not just planting trees out here. We're getting these kids off the street. We're giving them jobs, giving them bus fare, getting them on a straight path out of here."

"If you take away anything from today, what you gotta know is that we're not just planting trees," she repeated. "We're saving lives."

Urban Forestry as a Solution

Many of the residents of West Oakland, CA, have never known anything but poverty. The median household income in West Oakland in 2000 was $17,945 compared to the larger city of Oakland at $39,626, and San Francisco's median income of $55,221. Only 19% of students at McClymonds (West Oakland) High School were reading at grade level, and only 5% eligible to go to a University of California school (US Census 2000). The community is hemmed in by freeways on all sides: a four-sided, particulate-spewing physical and emotional barrier to achievement.

So with all the social problems plaguing West Oakland, why would Kemba choose to start an urban forestry organization? What connection does the urban environment have to the health and wellbeing of its residents?

A healthy urban forest will improve air quality, lower temperatures, and provide much-needed shade. The social benefits of city trees are well documented, credited for raising property values, improving civic ties in neighborhoods, boosting public health, encouraging visitors from out of town, and even increasing neighborhood safety (Sullivan and Kuo 1996; Kuo and Sullivan 2001; Wolf 2004, 2007; Wolf and Bratton 2006). Research on the human impact of urban trees and vegetation has shown that people are happier and healthier when they live in a greener environment (Taylor et al. 2001, 2002). Trees control local temperature and humidity, prevent erosion, absorb noise, and mitigate air pollution (Lull and Sopper 1969; Heisler 1986; Harris 1992; Nowak 1994). Compared to a treeless landscape, downtown shoppers linger longer, traffic slows, crime is reduced, and property values soar where a healthy tree canopy shades crowded urban streets (Sullivan and Kuo 1996; Kuo and Sullivan 2001; Wolf 2003, 2004, 2007; Wolf and Bratton 2006). Trees and natural views have been found favorable in the treatment of Attention Deficit Disorder, and in an area where about one third of households are below the poverty line, every job created to plant and maintain those trees can be a significant source of income (Taylor et al. 2001).

City trees are also part of an urban ecosystem that includes shrubs, grass, gardens, birds, insects, and—those oft-maligned species—humans. Often an urban forest ecosystem is referred to as green infrastructure, the chlorophyll and carbon composed objects in a city that create a healthful habitat and connect it via trails, streams, and other pathways to nature outside the city limits. Trees working as part of this green infrastructure produce oxygen, clean pollutants out of the air we breathe, filter stormwater, and provide habitat for diverse species. The urban forest, like any other "natural" or rural forest, needs to be managed for the provision of these benefits. And it is through this active management—engagement—that people become connected to and benefit socially and economically from the urban forest.

To Kemba, these benefits are real and much needed in West Oakland. Since incorporating Urban Releaf in 1998, she and her staff have planted close to 14,000 trees, 8,000 of which are fruit trees (Shakur 2009). Kemba has also been the recipient of numerous awards, including the Jefferson Award for Public Service in 2005. She built Urban Releaf from nothing, relying on community meetings, flyers, word-of-mouth, and converting community skeptics with tangible results. It's these results—well established by urban forestry scientists—that will lead to incremental, but durable changes in the quality of life for West Oakland residents.

The Science behind Urban Forestry

Urban Heat Island Effect

Phoenix, Arizona, is a smolderingly-hot city located in the northern portion of the Sonoran Desert. The average temperature in downtown Phoenix in the middle of July is 105 degrees Fahrenheit; in January, the average temperature sinks to a relatively chilly 66 degrees. Most of the homes located in upper-income neighborhoods are equipped with centralized air conditioning to deal with the summer heat, and many also have backyard swimming pools. However, the poorer neighborhoods, many of which are dominated by minority residents, are some of the hottest areas in the city and the least equipped to deal with the heat. To combat heat stress, many inhabitants sleep outside at night and rely on the Fire Department to bring cases of water (Larsen 2007).

The city of Phoenix records even hotter temperatures than the surrounding desert areas because of an environmental phenomenon called the Urban Heat Island Effect. The heat signature of the city reads like an island sticking out of a sea of average temperatures. A heat island is created when certain features of cities like asphalt, a lack of vegetation, or increased air pollution trap heat within the city. The higher temperatures can lead to greater concentrations of smog, increased human discomfort and disease. These effects can in turn drive increased demand for water and cooling, resulting in increased energy usage and carbon dioxide emissions, exacerbating the problem (McPherson 1994; McPherson and Simpson 2001). Heat islands are also known to raise city service expenditures and associated costs. Scientists at the Lawrence Berkeley Laboratory estimate that total national costs for offsetting the effects of increased urban temperatures during the summer are about $1 million per hour or over $1 billion per year (McPherson 1994).

The communities most affected by increased urban temperatures are generally poorer and composed predominantly of people of color. University of Michigan Urban and Regional Planning professor Larissa Larsen has studied the Phoenix urban heat island, especially the relationship between the physical landscape and its social context. She has found reciprocal relationships between exposure to heat stress and percentages of poor and minority inhabitants. The homes of lower-income inhabitants may have asphalt roofing—a low albedo, heat-retaining material—while higher-income residents use tile—a higher albedo, heat-reflecting material. Poorer residents may also have thinner walls incapable of keeping air conditioning indoors, or may not be able to afford air conditioning at all (Iverson and Cook 2000; Harlan et al. 2006; Jenerette et al. 2007). Wealthier communities not only have a better engineered environment, they also have a healthier natural environment. They maintain neighborhood parks and greenery, though often through the creation of an exclusive homeowners association, effectively privatizing greenspace and its benefits.

"The idea [of a homeowners association] is that the city has limited money to spend on its parks," Larsen said during a phone interview. "But by using homeowners associations, people get away from city management. You have wealthy communities that pay a fee every month, the money goes into their kitty, and then they can spend it for maintenance of these spaces. In poorer areas, they don't have as much money, and they also tend to be older areas that don't have homeowners associations. [The associations] are spreading throughout the country as municipalities see this as a way to avoid public expense. More and more developments in all parts of the country are engaging in them too, but it's huge in the Southwest" (Larsen 2007).

While increasing vegetation can provide temperature relief, urban forestry is by no means a universal panacea. The most important environmental issue facing Phoenix is access to water. Because trees need water to survive, tree-planting in Phoenix is a mixed blessing. In the past decade, xeriscaping—landscaping using native drought-tolerant plants—has become a popular alternative for homeowners conscious about the environment or high water bills.

Air quality

Poor air quality has been cited as one of the indicators of excessive environmental burden in a community. Environmental justice literature has helped to establish a link between proximity to pollution sources, health disparities, race, and income (Mohai and Bryant 1987). Many of these findings suggest that people of color and people in lower-income communities are more likely to bear environmental burdens and suffer from toxin-related health problems than whites or people in wealthier communities (Mohai and Bryant 1992; Ecob and Smith 1999). Scorecard.org, an organization devoted to ranking cities according to their "environmental toxicity," ranks West Oakland in the 70th percentile as one of the "Dirtiest Counties in the US" in all categories, including cancer risk, noncancer risk, carcinogens, developmental toxicants, and reproductive toxicants (Scorecard 2009).

On the opposite coast, Million Trees NYC chose East Harlem as a target neighborhood in its "Trees for Public Health" Program because of its lower-than-average level of street trees and higher-than-average asthma rates (Rosen and Greenfeld 2006). Most pollutants are attributed to mobile sources like cars, trucks, airplanes, ships, and construction. One major offender, city bus depots, may be responsible for the excessive asthma rates among many young people of color in New York City. The city is among the top four for increasing asthma deaths each year among individuals between the ages of 5 and 34 (Kinney 2000). Deaths from asthma between 1982 and 1987 came to an average annual rate of 1.2 per 100,000 (Carr et al. 1992). Of these deaths, 76.2% were among Blacks and Hispanics (Carr et al. 1992). Over eighty percent of all asthma hospitalizations in New York were individuals of either Black or Hispanic racial groups between 1982 and 1986 (Carr et al. 1992). The asthma rate for Whites was about 12 per 10,000 people, while rates for Blacks and Hispanics were right around 60 per 10,000—five times greater than that for Whites. However, the most shocking number appears at a smaller scale, in East Harlem, where 93% of residents are Black or Hispanic and asthma hospitalization occurs at a rate of 115 per 10,000 residents. In East Harlem, many of New York City's buses come to rest each night in enormous depots where miniscule particulate matter from diesel engines contributes to the incidence of asthma nearby. Their continued presence and a lack of pollution mitigation were strongly contested by environmental watch dog groups like West Harlem Environmental Action, Inc., or WE ACT. Now, WE ACT is in discussions with the Metropolitan Transportation Authority (MTA) to retrofit or close some uptown bus depots (WE ACT 2009).

Bronx County also has more than its share of asthma cases. Very near East Harlem, Mott Haven and Hunts Point hospitalization rates due to asthma for children in the South Bronx is 23.2 per 1,000 children, nearly 140 percent higher than New York City's rate of 9.9 per 1,000 children (Maantay 2007). Like the highly-concentrated bus depots in East Harlem, it is the Bronx's numerous busy highways connecting New York City's other boroughs to the mainland that are partly to blame for poor air quality (Maantay 2007).

Some species of urban trees have been found to adsorb—or collect on their leaf and bark surfaces—air pollutants including diesel particulates, cleansing the air while providing a multitude of other social and environmental services at the same time (Harris 1992; Nowak 1994). Unfortunately, several canopy studies done in cities have shown that in lower income

77

communities and communities of color, urban tree canopy is likely to be scarce, particularly in comparison to that of affluent and White-dominated areas of town (Million Trees LA 2009).

Crime prevention: hard evidence for the softer side of trees

In addition to health benefits like cleaner and cooler air, there is a general human understanding that people like trees. Of course, this is not without exception, and used to be based solely on anecdote or our "gut feeling" that nature is soothing, beautiful, and healthful. As social science progresses, however, data show that the effect of nature's intangible benefits can be measured and that our "guts" were right all along. Trees provide a number of social benefits, and one of the most shocking, potentially life-saving off-shoots of a healthy urban forest, is a reduction in violent crime.

Francis Kuo and William Sullivan, researchers at the University of Illinois at Urbana-Champaign, studied crime statistics for residents of Chicago's Ida B. Wells housing project in relation to their urban environment (Kuo and Sullivan 2001). What Kuo and Sullivan (2001) needed for their study was a housing complex with enough variation in vegetation to register a significant difference and enough similarity among the participants to determine whether vegetation was the only factor influencing their behavior. This particular development offered a full spectrum of housing environments with sufficient similarity that Kuo and Sullivan (2001) were able to isolate vegetation as the element affecting crime statistics around each apartment building. They found, conclusively, that vegetation was negatively correlated with crime in the area they studied. Using data for both property crime and violent crime, Kuo and Sullivan (2001) found that crimes reported per building fell with increasing amounts of vegetation. Areas with a medium amount of vegetation reported a 42% decrease in total crime versus areas with a low amount of vegetation. Areas with a high amount of vegetation reported a 52% decrease in total crime versus those same low vegetation areas.

Kuo and Sullivan's (2001) findings not only uncovered a negative correlation between vegetation and crime, but found that when traditional confounding variables for environment and crime were taken into account, vegetation adds an element of predictability previously unheard of. Vacancy rates, building height, number of apartment units, and number of occupied apartments per building are the top four traditional confounds. When compared, the best predictor of crime statistics was the combined use of vegetation and the number of units per building (Kuo and Sullivan 2001), an innovation that should spark action in public housing administration. Whatever else is going on in a public housing environment, we now know if we plant more trees and reduce the number of people living in very close quarters, we can reduce violent crime.

Apart from outside crime, a reduction in domestic violence is also associated with more trees in the urban environment (Sullivan and Kuo 1996). By providing a comfortable place to be outside of their small apartments, trees and other vegetation help establish community among the residents of public housing. The strains of poverty and the stress of daily life can be insurmountable when faced in isolation, but a community built with a pleasant outdoor space enables socialization and encourages people to rely on one another for support and solutions. Residents of buildings with more tree cover report constructive, non-violent conflict resolution methods with their spouses and children (Sullivan and Kuo 1996). As neighbors turn to one another for support, the individual stresses of coping with poverty, violence, and raising children seem to diminish and the demand on already strained municipal social services is lessened (Sullivan and Kuo 1996; Kuo 2001).

Furthermore, the very act of planting and caring for the urban forest can provide an excuse for neighborhood interactions and trust-building. When Kemba rolled up in her dark van next to the three teenage boys on the sidewalk in West Oakland, she did so slowly, rolling down my passenger side window on the approach, straining across the car to stick her neck out—literally—to show the kids she wasn't a threat. She spoke to them like she would speak to her own children. No fear, no pretense. No violence.

"It's dangerous out there," Kemba said afterwards, reflecting on the experience. "We've gotta let these neighborhood kids know who we are. We're already an oddity—these people aren't used to seeing people of color planting trees. And we're in a big dark van, driving up and down these streets. When they see us, we want them to know what we're doing. And we want them to help us."

Environmental Justice and Urban Tree Canopy Studies: Analyzing Urban Forests, Planning for Healthy Ones

The next frontiers in urban forestry analysis are canopy studies and tree censuses, which have been conducted in several cities nationwide. In order to improve the health and complexity of the urban forest, scientists must understand its current extent as well as planting opportunities and challenges. A common finding is that when urban tree canopy is mapped across a city, low canopy often coincides with low income (Million Trees LA 2009). This mapping technology provides a way to look at a city from above and determine where to focus tree planting efforts.

Tree canopy studies help all the main players in an urban forestry initiative to see the total picture. Canopy maps are literally city-wide maps with canopy gradations overlain in percentages; the darker the green, the more robust the urban forest. The maps and the research that precedes them allow the city and civic partners to work from the same body of information and set goals that complement one another.

Several cities have initiated efforts to plant one million trees in their city parks, in school yards, along streets, and on privately-owned land. The City of Los Angeles has benefited from a close relationship with the U.S. Forest Service, which conducted research on the city's urban tree canopy for the Million Trees LA campaign. The study proved that the city has room for a million trees and showed widely varying tree canopy cover in different parts of the city (McPherson et al. 2008). However, with such a daunting task as planting one million trees, the campaign itself would have a hard time engaging the community in the effort.

While Million Trees LA set about focusing on low-canopy areas, one of many local non-profit organizations they turned to was TreePeople. TreePeople chose to partner with the LA Department of Recreation and Parks to help carry out this bold initiative. Its role involves training and supporting volunteers to plant and care for trees in parks, on school campuses and along city streets (Million Trees LA 2009). Their focus on low-canopy, low-income areas is part of a growing trend to target urban forestry efforts in the areas of greatest need.

Another community based organization, Casey Trees in Washington, D.C., took the canopy study a step further and issued a city "report card" (Casey Trees 2010). The report card grades the District of Columbia on five main criteria: tree coverage, health, planting, protection, and awareness, and gave the city an overall grade of 'B.' In order to accomplish such a comprehensive look at tree health, street tree and park planting, ordinances, and advocacy efforts, Casey Trees reached out to numerous partners and their own volunteers to get the information they needed. The Tree Report Card is laid out to be easily replicated in any other city, and it presents an opportunity to include other factors like race, poverty, crime, and public health. It is a nice look ahead to the melding of physical and social science in the advancement of urban forestry as a discipline and another tool in the urban solutions toolbox.

Tree-Planting: How it Happens

During the past 50 years, thousands of organizations like Kemba's have sprung up around the United States. TreePeople, a community-based organization in Los Angeles, was one of the first. These local non-profits are supported by community donations, other neighborhood organizations, city and state government, corporations and private foundations, and federal entities like the U.S. Forest Service. Support comes in the form of financial, technical, and even physical (the kind with a pick-axe and shovel) assistance.

CBOs: The business of community-building

The South Bronx is a predominantly minority community with a history of high crime and poverty. In the past few years, the community has changed both physically and demographically. Encroaching gentrification has led some locals to begin affectionately referring to the area as "SoBro," a play on the trendy and gentrified SoHo in Manhattan. The area also appears significantly greener than it used to, in no small part due to the work of MacArthur Award-winning activist Majora Carter and her organization, Sustainable South Bronx. With the help of a passionate staff, Majora turned Sustainable South Bronx from a struggling nonprofit into one of the pre-eminent environmental justice organizations in the nation. One of her greatest accomplishments is the creation of Hunts Point Riverside Park near the Bronx River, built on top of a site that used to be a landfill. The organization recently organized the first annual "Hunts

Point Hustle," a 5K race to encourage local residents to run, walk, or dance through the streets of Hunts Point. A mysterious streak of neon-green paint bisects many of the sidewalks—a charming touch of clandestine environmental activism—and leads to the newly-developed Hunts Point Riverside Park. The green line is to the South Bronx environmental community what an appearance on SportsCenter is to a professional athletes: an arrival in style.

Elena Conte, the Solid Waste and Energy Coordinator with Sustainable South Bronx in 2007, took me on a tour down the Living Memorial Trail, which runs along Lafayette Avenue to the Hunts Point Riverside Park. The trail consists of street trees, perennial plantings, decorated tree guards, and an expansive mural on building walls to depict the past, present, and future of Hunts Point. Lafayette Avenue used to be one of the most run-down streets in the neighborhood, known for its broken sidewalks and dilapidated buildings. Now, Lafayette is one of the greenest streets in town. Elena was both intimately familiar with the inner workings of the project as well as the surrounding neighborhood. Like Kemba, she seemed to be fluent in the culture of the community, flipping easily between Spanish and English, technical ecological language and passionate social advocacy. She seemed to know everyone she passed, stopping frequently to sit down and chat, congratulate folks on a recent wedding or childbirth, or simply to ask whether they liked the newly planted trees.

Walking back up Lafayette Avenue, I reflected on Elena's and Kemba's accomplishments, and the respect they each had earned from their communities. They had helped to transform their organizations into thriving centers of their respective communities. They had done so by understanding the culture and histories of their neighborhoods, by teaching the virtues of urban forestry, and by giving locals the tools to both plant trees and save lives.

Three-thousand miles across the country from the South Bronx, TreePeople founder, president, and charismatic leader Andy Lipkis pioneered the art of urban forestry. TreePeople has worked since 1973 to "help nature heal our cities" throughout the Los Angeles region. By creating the "Citizen Forester" model with his wife Katie, Andy put the power of change in the hands of Angelinos. They come to free trainings to learn how to organize small "green teams," obtain proper permits, recruit volunteers, track down grant money, plant trees, and care for the urban forest. TreePeople staff provide assistance throughout the process, including trucks and tools. The Citizen Forester model provides a framework for self-empowerment through urban forestry.

The Urban Resources Initiative (URI) operates on the same principle; they too have a charismatic leader. URI's Greenspace Manager, Chris Ozyck, has become, albeit accidentally, something of a small-town celebrity; but it's the community that leads local planting efforts. By offering mini "grants" to neighborhood groups in New Haven, CT, URI provides the dirt, trees, and tools to bring the neighborhoods' ideas to life. Instead of providing funding for street tree projects, pocket parks, or other public space improvement, URI brings the material to supplement neighborhood dreams, planning, and labor.

In some communities, summer street tree plantings have led to Block Watch groups, as people who never met before start chatting about other changes they would like to see in their neighborhoods. One evening or weekend afternoon each week during the summer, neighbors come out with their kids and friends to plant trees and spread mulch around last year's plantings. Each week, new faces come and go, spreading the word about their community-initiated work. Group size may wax and wane, but the core group remains and for some neighborhoods this has gone on for fifteen years. These groups are the guiding force behind urban forestry in New Haven.

Governments

Some of the most important partners in urban forestry are city, state, and federal urban forestry agencies. Often the silent, supportive partner, the U.S. Forest Service (USFS) Urban and Community Forestry (UCF) program makes possible a great deal of work by state and local entities, as well as non-profit organizations, through grants and technical assistance.

The beginnings of UCF can be traced to back to Dutch Elm Disease, a devastating fungus spread by the elm bark beetle. By the mid-twentieth century, the disease had killed trees across the United States, particularly in the Midwest where cities and towns had planted Elm monocultures along their streets. The epidemic spread from the Eastern Seaboard west to the Rocky Mountain states. Prior to this disease, most small cities and towns did not have a tree program, and the cities that did have an arboriculture program were mostly focused on managing single tree species, not urban forest ecosystems.

In 1970, a USFS pilot program was started in Colorado to fund a technical specialist to work with communities affected by Dutch Elm Disease. A flurry of reports by various commissions followed, suggesting to Congress that foresters needed to become more sympathetic to an increasingly urban population, and that an urban and community forestry program would be a wise investment. In 1978, Congress officially authorized the creation of the Urban and Community Forestry Program via the Cooperative Forestry Assistance Act (CFAA). The Act cited findings that forested land and associated natural resources enhance economic value of commercial and residential areas; that urban forests were crucial in the fight against the urban heat island effect, rising carbon dioxide levels, and rising energy costs; and that the health of urban forests was on the decline. Congress authorized the program to provide technical assistance as well as competitive, matching grants to local governments and nonprofit organizations—each federal dollar would be matched with a dollar from the recipient or a project partner. The federal UCF program operates with a slim national staff in Washington, D.C., focused on national policy and budget issues, and several regional program managers responsible for the care and feeding of state and local programs. All told, there are fewer than 20 federal UCF employees across the nation. States, in order to receive federal funding, identify an urban and community forestry coordinator, volunteer coordinator, and an urban forestry council. This model is repeated at the city level, the most highly functioning programs supporting a city urban

82

forester and a local council or tree board taking responsibility for working with the community and community-based organizations to plant and maintain the urban forest.

Struggle for Legitimacy

Despite the momentum building in favor of urban forestry at the beginning of the twenty-first century, urban forestry is not viewed as a top agency or national priority and is criticized for not being "real" forestry. Why, with over 80 percent of the population living in urban areas, is it so difficult to establish urban forestry as a scientific discipline and a crucial federal program? Even President Obama's White House Office of Urban Affairs, which policy-minded urban foresters viewed with initial excitement, does not involve an urban forester. The Urban White House webpage states that, "The President's urban agenda will promote cross-cutting plans to revitalize urban areas, considering housing, transportation, energy, labor, education, and criminal justice policy as a system rather than independent of each other" (WHOUA 2010). While urban forestry can have significant positive impact on all these issues, the discipline was not invited to the table and was not included in the otherwise robust systems approach to urban revitalization.

Another national opportunity for urban forestry arises out of First Lady Michelle Obama's "Let's Move!" anti-obesity initiative. While in its infancy at the time of this writing, one pillar of the initiative is physical activity (Let's Move 2010). A healthy urban forest is critical to getting kids safely outside by providing slower traffic, shady streets, and welcoming play areas.

Land management agencies as a whole appear to be struggling to maintain their relevance with an increasingly urbanized public. The rise of the YouTube, Google-eyed generation, and a more sedentary public are more indicators that our natural landscape has declined in importance in the public eye since the early 1900s, which saw the creation of the National Park Service and U.S. Forest Service to preserve and manage the nation's forests and natural resources. Why visit the Great Outdoors when, with the use of a Nintendo Wii or Sony PlayStation, a semblance of the outdoors can be brought into the comfort of one's own living room?

Urban foresters must also contend with a deep bias within the environmental community against the value of cities. This bias can be partially attributed to environmental writers like Edward Abbey, who often wrote about his connection with the "natural" world at the expense of his relationship with urban life. In his novel <u>Desert Solitaire</u>, Abbey writes, "How difficult to imagine this place without a human presence; how necessary. I am almost prepared to believe that this sweet virginal primitive land will be grateful for my departure and the absence of the tourists, will breathe metaphorically a collective sigh of relief—like a whisper of wind—when we are all and finally gone and the place and its creations can return to their ancient procedures unobserved and undisturbed by the busy, anxious, brooding consciousness of man" (Abbey 1968).

Admittedly, Abbey is one of the more notorious environmental writers, known as much for his exquisite descriptions of desert wilderness as his stinging, mischievous wit. Nonetheless, the above paragraph contains a key to understanding the mentality of the twentieth century environmentalist. What Abbey calls the "sweet virginal primitive landscape," is nature that is untouched, unused, and perhaps even unseen by humans. Simply by the existence of their "brooding consciousness," humans taint the wilderness—they force it to hold its breath, halt its natural processes, force it to suffocate and wither into an infected version of its former self. Cities, therefore, could be considered a raw mass of held breath and stopped processes. They are spoiled, corrupted, the impure mistress in an environmentalist's marriage to the Natural.

But cities are habitats. They have become the primary habitats for humans around the world, as well as for many species of trees, insects, and animals. We breathe city air, we drink city water. If we continue to think of humans as harbingers of impurity, then we're planting seeds for the adulteration of our planet. If we embrace cities as living, breathing, ecological entities, as homes for urban forests, then we will improve our quality of life and succeed in conserving planetary resources for future generations. Urban forestry is as much about saving the planet as it is about saving communities, and believing in our own abilities to make lasting social change.

"The way I started," Kemba remembered, "was at 29th and West—and I was standing there with a lady, and she said, 'We need trees.' So I called the city, and the city said it would take a year to get a tree. Remember, this block had two trees. I'm like, 'A YEAR? You SERIOUS?' So I called California Releaf and Friends of the Urban Forest. They told me, 'Just do it Kemba! Get a grant!' So I got a small grant for $4,000 or $5,000 that planted about thirty trees. Once I did that, I thought, 'This is nothing.'" Kemba laughed. "We need more than thirty trees."

References

Abbey, E., 1968. Desert Solitaire. Touchstone, New York.

Carr, W., L. Zeitel, and K. Weiss, 1992. Variations in asthma hospitalizations and deaths in New York City. American Journal of Public Health. v. 82, pp. 59-65.

Casey Trees, 2010. First Annual Tree Report Card. Available online at http://www.caseytrees.org/geographic/key-findings-data-resources/tree-report-card/index.php.

Ecob, R. and G. Davey Smith. 1999. Income and health: What is the nature of the relationship? Social Science and Medicine, v. 48, pp. 693-705.

Harlan, S. L., A. J. Brazel, L. Prashad, W. L. Stefanov, and L. Larsen, 2006. Neighborhood microclimates and vulnerability to heat stress. Social Science and Medicine, v. 63, p. 2847-2863.

Harris, R.W. 1992. Arboriculture: Integrated Management of Landscape Trees, Shrubs, and Vines. Prentice Hall, Englewood Cliffs, NJ.

Heisler, G. M., 1986. Energy savings with trees. Journal of Arboriculture. v. 12, pp. 113-125.

Iverson, L. R. and E. A. Cook, 2000. Urban forest cover of the Chicago region and its relation to household density and income. Urban Ecosystems v. 4, pp. 105–124.

Jenerette, G. D., S. L. Harlan, A. Brazel, N. Jones, L. Larsen, and W. L. Stefanov, 2007. Regional relationships between surface temperature, vegetation, and human settlement in a rapidly urbanizing ecosystem. Landscape Ecology, v. 22, p. 353-365.

Kinney, P. L. et al., 2000. Airborne concentrations of PM2.5 and diesel exhaust particles on Harlem sidewalks: A community-based pilot study. Environmental Health Perspectives, v. 108, pp. 213-218.

Kuo, F. E., 2001. Coping with poverty: Impacts of environment and attention in the inner city. Environment and Behavior, v. 33, pp. 5-34.

Kuo, F. E. and W. C. Sullivan, 2001. Environment and crime in the inner city: Does vegetation reduce crime? Environment and Behavior, v. 33, pp. 343-367.

Larsen, L., 2007. Personal Communication with Sarah Levy.

Let's Move! 2010. America's Move to Raise a Healthier Generation of Kids. Available online at http://www.letsmove.gov/.

Lull, H.W. and W. E. Sopper, 1969. Hydrologic effects from urbanization on forested watersheds in the Northeast. USDA Forest Service, Research Paper, NE-146, pp. 1-31.

Maantay, J., 2007. Asthma and air pollution in the Bronx: Methodological and data considerations in using GIS for environmental justice and health research. Health and Place, v. 13, pp. 32-56.

McPherson, E. G., 1994. Cooling urban heat islands with sustainable landscapes. In The Ecological City: Preserving and Restoring Urban Biodiversity, H. P. Rutherford, R. A. Rowntree and P. C. Muick (eds.). University of Massachusetts Press, Amherst, MA, pp. 151-171.

McPherson, E. G. and J. R. Simpson, 2001. Effects of California's Urban Forests on Energy Use and Potential Savings from Large-Scale Tree Planting. USDA Forest Service, Pacific Southwest Research Station, Center for Urban Forest Research, Davis, CA.

McPherson, E. G., J. R. Simpson, Q. Xiao, and W. Chunxia, 2008. Los Angeles 1-Million tree Canopy Cover Assessment. General Technical Report PSW-GTR-207. USDA Forest Service, Pacific Southwest Research Station, Albany, CA.

Million Trees LA, 2009. Million Trees LA. Available online at http://www.milliontreesla.org.

Mohai, P. and B. Bryant, 1987. Toxic Wastes and Race in the U.S.: A National Report on the Racial and Socio-economic Characteristics of Communities with Hazardous Waste Sites. Commission for Racial Justice, United Church of Christ.

Mohai, P. and B. Bryant (eds.), 1992. Environmental Racism: Reviewing the Evidence. In Race and the Incidence of Environmental Hazards: A Time for Discourse. Westview, Boulder, CO.

Nowak, D. J., 1994. Air pollution removal by Chicago's urban forest. In Chicago's Urban Forest Ecosystem: Results to the Chicago Urban Forest Climate Project, E. G. McPherson, D. J. Nowak, and R.

A. Rowntree (eds.). General Technical Report NE-186. USDA Forest Service, Northeastern Forest Experiment Station, Radnor, PA, pp. 63-81.

Rosen, L., and J. Greenfeld, 2006. East Harlem: A Community Forestry Management Plan. City of New York Parks and Recreation.

Scorecard, 2009. Pollution in your Community. Available online at http://scorecard.goodguide.com/.

Shakur, K., 2007, 2009. Personal Communication with Sarah Levy.

Sullivan, W. F. and F. E. Kuo, 1996. Do Trees Strengthen Urban Communities, Reduce Domestic Violence? Urban and Community Forestry Assistance Program, Technology Bulletin.

Taylor, A. F., F. E. Kuo, and W. C. Sullivan, 2002. Views of nature and self-discipline: Evidence from inner city children. Journal of Environmental Psychology. v. 22, pp. 49-63.

Taylor, A.F., F. E. Kuo, and W. C. Sullivan, 2001. Coping with ADD: The surprising connection to green play settings, Environment and Behavior, v. 33, pp. 54-77.

WHOUA (White House Office of Urban Affairs), 2010. Urban Policy. Available online at http://www.whitehouse.gov/issues/urban-policy.

US Census Bureau, 2000. United States Census 2000, American Fact Finder. Available online at http://www.census.gov/main/www/cen2000.html.

WE ACT, 2009. Achieving Environmental Justice by Building Healthy Communities Since 1988. Available online at http://www.weact.org.

Wolf, K. L., 2003. Public response to the urban forest in inner-city business districts. Journal of Arboriculture, v. 29, pp. 117-126.

Wolf, K. L., 2004. Trees in Small City Business Districts: Comparing Resident and Visitor Response. Human Dimensions of the Urban Forest, Fact Sheet #16.

Wolf, K. L., 2007. City trees and property values. Arborist News, v. 16, pp. 34-36.

Wolf, K. L. and N. Bratton, 2006. Urban trees and traffic safety: Considering US roadside policy and crash data. Arboriculture and Urban Forestry, v. 32, pp. 170-179.

Chapter 5: Self-organizing Systems and Environmental Justice: Application to Arsenic Contamination of Groundwater in Nepal

Steven H. Emerman, Aimee J. Luhrs, Susan E. Sandford and Adam Finken

Summary

The contamination of water sources has always disproportionally affected the poor as those least able to purchase clean drinking water or afford the construction of community water treatment facilities. In many cases, environmental problems that disproportionally affect the poor have not been solved simply because resources have not been expended toward their solution. In other cases, there is a dominant scientific paradigm that does not permit any solution, while there is an alternative scientific paradigm within which a solution could become possible. Often the dominant scientific paradigm that does not permit solutions to the environmental problems that affect the poor is intimately connected with a dominant social paradigm that does not recognize the rights of the poor or the connections between all people. This chapter will address the problem of arsenic contamination of groundwater in Nepal as an issue of environmental justice. The dominant scientific paradigm with regard to arsenic is that elevated levels of arsenic in South Asia are due to the strongly reducing (low oxygen) conditions in the flood plain of the Ganges River and are unrelated to any human activities. In the field of ecology there is an alternative paradigm that, in some ways, an ecosystem mimics the behavior of a single organism. We believe that this alternative paradigm of ecology has not been widely accepted by the scientific community because of the connections it draws among all living things, which can lead to uncomfortable conclusions regarding the connections among all people. However, it follows logically from this alternative paradigm that the arsenic contamination of groundwater is, in fact, related to human activity and is linked to the loss of healthy forests and grasslands in Nepal.

Introduction

The need for clean drinking water is the most fundamental of all human needs. The contamination of water sources has always disproportionally affected the poor as those least able to purchase clean drinking water or afford the construction of community water treatment facilities. In many cases, environmental problems that disproportionally affect the poor have not been solved simply because resources have not been expended toward their solution. In other cases, there is a dominant scientific paradigm that does not permit any solution, while there is an alternative scientific paradigm within which a solution could become possible. Often the dominant scientific paradigm that does not permit solutions to the environmental problems that affect the poor is intimately connected with a dominant social paradigm that does not recognize the rights of the poor or the connections between all people.

This chapter will address the problem of arsenic contamination of groundwater in Nepal as an issue of environmental justice. The dominant scientific paradigm with regard to arsenic is that elevated levels of arsenic in South Asia are due to the strongly reducing (low oxygen) conditions in the flood plain of the Ganges River and are unrelated to any human activities. In the field of ecology there is an alternative paradigm that, in some ways, an ecosystem mimics the behavior of a single organism. This alternative paradigm is simply the application of complexity theory to ecology. The related Gaia Hypothesis, that the entire Earth, in some ways, mimics the behavior of a single organism, is the application of complexity theory to the entire Earth. It will be shown in this chapter that it follows logically from this alternative paradigm that the arsenic contamination of groundwater is in fact, related to human activity and is linked to the loss of healthy forests and grasslands in Nepal. Without an understanding of the connections involved in arsenic contamination, it will not be possible to seek effective solutions.

Although there is a great deal of evidence in favor of the alternative paradigm in ecology (see references throughout this chapter), the alternative paradigm, and its global equivalent called the Gaia Hypothesis, have been only marginally accepted in the scientific community. We believe that this alternative paradigm of ecology has not been widely accepted by the scientific community because of the connections it draws among all living things, which can lead to uncomfortable conclusions regarding the connections among all people. The dominant ideologies of modern Western civilization are nationalism and capitalism. The key aspects of these interconnected ideologies are respectively:

1) The natural order of the world is the division of people into nations. Although there are some mutual obligations among the citizens of a nation, what a citizen of one nation owes to a citizen of another nation is extremely limited.
2) The natural order of the world is that the obligations of one person toward another are based upon the other's ability to pay or otherwise offer something in exchange. Even among the citizens of a nation, what the rich owe to the poor in the absence of adequate payment is extremely limited.
3) The alternative paradigm, which emphasizes the intimate interconnections among all living things and their nonliving environment, cannot be entertained without seriously undermining the dominant ideologies, which most (but not all) scientists are not yet ready to do.

Another possible explanation for the marginal acceptance of the alternative paradigm and the Gaia Hypothesis is that these ideas regarding interconnections among rocks, animals, plants, air and water cannot be comprehended without a knowledge of a wide variety of scientific disciplines, including geology, microbiology, soil science, oceanography, meteorology, and many other fields. Not many geologists have studied microbiology. Few microbiologists are qualified in meteorology. There are interdisciplinary fields such as geomicrobiology, but thus far, they tend to not be global in scope. We do not believe that this non-ideological argument is an adequate explanation for the marginalization of the alternative paradigm and the Gaia Hypothesis. Geologists were willing or were forced to learn physics with the discovery of the theory of plate tectonics. Biologists became chemists and biochemists after the discovery of

DNA. However, neither plate tectonics nor DNA offers such a profound attack on the nationalist-capitalist ideology of modern Western culture as does the Gaia Hypothesis.

This chapter will proceed in the following order:

1) We will review the problem of arsenic contamination of groundwater in Nepal and the development of the dominant scientific paradigm.
2) We will review the problem of deforestation in Nepal, which is normally regarded as unconnected with arsenic contamination of groundwater.
3) We will review the alternative paradigm in ecology.
4) We will review some basic ideas in soil science and vegetational succession.
5) We will use the alternative paradigm to make predictions regarding how changes in land use could affect the level of mobile soil arsenic (the soil arsenic that is available for leaching into groundwater).
6) We will review an experiment carried out in Nepal that tested the predictions made by the alternative paradigm.
7) We will describe an experiment carried out in Iowa that further tested the predictions made by the alternative paradigm. This experiment has not previously been described in the scientific literature.
8) We will use the predictions of the alternative paradigm and the results of experiments to make specific recommendations regarding how the problem of arsenic contamination of groundwater could be addressed in Nepal.

Arsenic Contamination of Groundwater in Nepal: The Dominant Paradigm

Over the past decade a great deal of attention has focused on arsenic contamination of groundwater in West Bengal and Bangladesh (Bhattacharaya et al. 1997; Dhar et al. 1997; Nickson et al. 1998) and in the Terai region (Indo-Gangetic plain) of Nepal (Bhattacharya et al. 2003; Emerman 2004, 2005; Kanel et al. 2005; Tandukar et al. 2005; Emerman et al. 2007, 2010, 2011). As of January 2004, 18,635 wells had been tested in the Terai region of Nepal, of which 23.7% exceeded the World Health Organization (WHO) guideline value of 10 µg/L, while 7.4% exceeded the Nepal Interim Standard of 50 µg/L (ENPHO and USGS 2004). There are many models for arsenic contamination of groundwater in South Asia, each one of which assigns responsibility to a different sphere of human activity or to no human activity at all. It is generally agreed that arsenic contamination is too widespread to be due to human activities such as smelting or use of arsenic-based pesticides (Aswathanarayana 1997). However, some workers have suggested that human activities have promoted the transfer of naturally occurring arsenic from sediment into groundwater. Badal et al. (1996) and Mallick and Rajgopal (1996) have argued that over-pumping of aquifers has caused oxidation of sulfide minerals and release of co-precipitated arsenic into groundwater. Acharyya et al. (1999, 2000) have proposed that excessive use of phosphate fertilizers has resulted in displacement of arsenic from sediment adsorption sites by phosphate. The most recent studies have argued that arsenic contamination is unrelated

89

to human activities. According to these studies, arsenic contamination results from the release of arsenic from adsorption sites on iron oxyhydroxides after dissolution of the iron oxyhydroxides (Nickson et al. 2000; McArthur et al. 2001; Bose and Sharma 2002; Harvey et al. 2002) or after reduction of adsorbed arsenic from arsenate (As^{+5}) to arsenite (As^{+3}) (Bose and Sharma 2002). Both processes are likely under the strongly reducing conditions found in the thick package of alluvial sediments in West Bengal, Bangladesh and the Terai region of Nepal (Bose and Sharma 2002). An alternative model is that arsenic is displaced from adsorption sites by carbonate after sediments deposited in surface waters with low carbonate concentration are later exposed to groundwater with high carbonate concentration (Appelo et al. 2002). Still another alternative is that arsenic is co-precipitated with diagenetic carbonate concretions and that arsenic is released into groundwater upon dissolution of the carbonate concretions under acidic conditions (Shanker et al. 2001). The iron oxyhydroxide reduction-dissolution model, which does not assign responsibility to human activity, is the dominant model at the present time. However, there are indications that the iron oxyhydroxide reduction-dissolution model, which was developed on the basis of data collected in West Bengal and Bangladesh, may not be applicable to Nepal. For example, there are elevated levels of arsenic in rivers draining into the Terai region of Nepal, which are not found in West Bengal or Bangladesh (Smedley and Kinniburgh 2002; Emerman 2005; Emerman et al., 2007, 2010, 2011).

Deforestation in the Terai Region of Nepal

In 1927, the Terai had 48% forest cover with 70% forest cover in western Terai. The remainder was largely covered with open grasslands of elephant grass and reeds with a few isolated small towns near the Indian railheads (Sharma 1995). Prior to 1950, the Rana government began cutting timber in the Terai for sale to British India (Sharma 1988). Following the overthrow of the Rana government in 1950, hill people started migrating to the Terai after the malaria eradication program of the 1950s. The deforestation of the Terai intensified in the 1970s due to the clearing of agricultural land and cutting of trees for timber and fuel wood. By 1987, only 10% of the Terai was covered by forest with only 4% forest cover in western Terai. As of 1994, the deforestation of all of Nepal was continuing at a rate of 2% reduction per year (Sharma 1995). The settlement of the Terai could be regarded as an unintentional experiment in the consequences of the rapid destruction of both forests and grasslands. Although grasslands normally expand when forests are cleared, in the case of the Terai, both forests and grasslands have been converted into either agricultural fields or intensively overgrazed pastures. There are virtually no healthy grasslands remaining in Nepal outside of Chitwan and Royal Bardia National Parks.

Although arsenic in water is regarded as toxic at levels exceeding 10 μg/L, the average arsenic concentration in unconsolidated sediment is 3 mg/kg with range 0.6-50 mg/kg (Smedley and Kinniburgh 2002). Drinkable groundwater results from the fact that, normally, the vast majority of arsenic is adsorbed on sediment or co-precipitated with sediment or exists in crystalline form in sediment and is not released into groundwater. The same discussion could be applied to soil. The average arsenic concentration in soil is 7.2 mg/kg with range 0.1-55 mg/kg (Smedley and Kinniburgh 2002). If even a small fraction of soil arsenic were mobilized and leached into groundwater, undrinkable groundwater could result. Therefore, it is difficult to ignore changes in land use, such as deforestation or overgrazing, in terms of understanding the arsenic cycle in South Asia.

Complexity Theory: The Alternative Paradigm in Ecology

Aldo Leopold (1949) wrote in A Sand County Almanac, "A science of land health needs, first of all, a base datum of normality, a picture of how healthy land maintains itself as an organism." The concept of land as an organism can now be understood in terms of the new field of complexity theory, also known as the theory of complex systems or the theory of self-organization. Complexity theory is the study of the properties that emerge when parts assemble into a whole (Kauffman 1993, 2000; Lewin 1993; Holland 1996; Coveney and Highfield 1996; Rayner 1997; Bak 1999; Bar-Yam 2003). An excellent example is consciousness, which is a property of a whole living organism, but which is not found in any of the cells, organs, or other parts of an organism. Complexity theory has been applied to a wide variety of fields but only recently to ecology and geology (Watson and Lovelock 1983; Klinger 1991; Jørgensen et al. 1992; Lovelock 1995; Solé and Manrubia 1995; Klinger and Short 1996; Klinger and Erickson 1997; Von Bloh et al. 1997; Levin 1998; Downing and Zvirinsky 1999; Harding 1999; Bradbury et al. 2000; Jørgensen and Müller 2000; Emerman and Parmelee 2002; Lenton and van Oijen 2002; Allen and Emerman 2003; Emerman 2004; Klinger 2004). (See Clements (1916), Margalef (1963) and Odum (1969) for historical antecedents of applications of complexity to ecology.) According to complexity theory, when the parts, such as plants, animals, microorganisms and soil, assemble into an ecosystem, properties emerge that allow the ecosystem, in some ways, to mimic the behavior of a single organism. (Another way to say the same thing is that homeostatic properties emerge that allow the system to self-regulate, much like an organism.) The Gaia Hypothesis proposes that the Earth's crust, atmosphere, oceans and living things can act as a unified whole that, in some ways, mimics the behavior of a single organism (Lovelock 1995). The Gaia Hypothesis is simply the application of complexity theory to the planet as a whole (Klinger 2004).

The Vegetational Succession

The vegetational succession is an old concept (Clements 1916), but an excellent series of papers by L. F. Klinger and his co-workers has contributed to a modern understanding of the vegetational succession (Klinger 1990, 1991, 1996a, 1996b, 2004; Klinger et al. 1990, 1994, 1996, 1998; Klinger and Short 1996). Any undisturbed landscape will undergo a succession from bare soil to annual grasses to perennial grasses to deciduous shrubs to deciduous forest to conifer forest to peatland. A natural or human-induced disturbance, such as landslide, tillage, excessive grazing, drought, fire or glaciation can reverse the succession and cause any landscape to revert to bare soil, whereby the succession begins anew. The proper combination of mild disturbances can halt the succession and maintain a landscape in a semi-permanent state of some form of vegetation. (Glacial cycles will prevent the existence of any truly permanent form of vegetation.) For example, the combination of periodic drought, occasional fire and mild grazing kept central North America in ecosystems dominated by perennial grasses (Anderson 2006). The objective of most agricultural practices is to maintain land in a semi-permanent state of annual grasses that does not revert to bare soil or progress to perennial grasses or shrubs. The vegetational succession can also be accelerated by increased moisture or environmental acidity. At the present time, on a global basis, grasslands are being invaded by shrubs and trees, while simultaneously, forests are in decline as trees are attacked by mosses and forest is replaced by peatland (Klinger 1990). Although the two simultaneous global processes seem paradoxical, they can both be understood as a shift to later stages of vegetational succession caused by decreased disturbance frequency and/or excessive environmental acidity.

The systematic classification of soils is an enormous topic (Brady and Weil 2007; Soil Survey Staff 2007). However, out of the 12 soil orders, for the purposes of understanding the argument of this chapter, we will consider only Mollisols, Inceptisols and Entisols. Mollisols typically develop under grasslands. They are characterized by a mollic epipedon, which is a soft, dark, thick (> 25 cm) mineral surface horizon. The dark color corresponds to high accumulated organic matter (> 0.6 % organic carbon), which results from the decomposition of grass roots. Inceptisols have less organic matter than Mollisols. They have only the inception of profile development. The surface horizon is either too thin or too light-colored to be regarded as a mollic epipedon. Entisols have the least organic matter of all the soil orders. Entisols show little, if any, profile development. They tend to develop when the soil is too frequently disturbed (for example, by drought, flooding or landslides) for soil horizons to develop. Although certain soil orders may tend to develop under certain types of vegetation, there is not a simple correspondence between soil order and present vegetation since the vegetational succession can proceed much faster than the development of soil horizons.

Changes in Land Use and the Mobilization of Soil Arsenic

We now combine complexity theory with the vegetational succession to develop five hypotheses regarding how changes in land use could result in the transformation of soil arsenic into mobile form, which could then be leached into groundwater. A self-regulating characteristic of single organisms that could be applied to ecosystems is the ability to sequester toxins. Arsenic is toxic to all plants and animals. According to the above reasoning, an ecosystem must include a microbial population that transforms arsenic into an immobile / non-bioavailable form such as a methylated form or an arsenosugar or arsenobetaine (Cullen and Reimer 1989; Gustafsson and Jacks 1995). As a landscape evolves through the stages of vegetational succession, the rate of floral growth decreases. Annual grasses must complete their life cycle in one season, while mosses (dominant organism of peatlands) grow much slower than vascular plants. In the early successional stages, a large microbial population whose function is to sequester arsenic could be a waste of energy, since plants can grow at a rate that keeps the carbon:arsenic ratio low. The related theories of complexity and vegetational succession predict that, as a landscape proceeds through the vegetational succession, an increasing proportion of soil arsenic will be found in immobile / non-bioavailable forms. This prediction could be tested by comparing total soil arsenic and mobile soil arsenic across ecosystem boundaries. However, the measurement of total arsenic involves hot concentrated acids and is a time-consuming process. In view of the high spatial heterogeneity of all soil properties and in the interest of measuring as many soil samples as possible, this study measured only mobile soil arsenic under the assumption that total soil arsenic will remain relatively constant. Therefore, the first preliminary hypothesis of this study is that, when mobile soil arsenic is compared across ecosystem boundaries, the earlier stage of vegetational succession will show higher mobile soil arsenic.

The significance of soil order and its level of organic matter is that, when soil organic matter is abundant, sufficient energy is available so that all ecosystems could include a microbial population that sequesters arsenic. Therefore, the first preliminary hypothesis can be refined to state that when soil mobile arsenic is compared across ecosystem boundaries in Entisols, the earlier successional stage will show higher mobile soil arsenic. However, mobile soil arsenic will not change across ecosystem boundaries in Mollisols. Although Inceptisols may be more chemically related to Entisols than to Mollisols, whether Inceptisols should be grouped with Entisols or with Mollisols in the context of the first preliminary hypothesis must be determined by experiment.

Another characteristic of organisms that can be applied to ecosystems is the ability to expel toxins. This characteristic is related to a key feature of complex systems, which is that they must possess both symmetry-building behaviors and symmetry-breaking behaviors. Symmetry refers to repeatability in space and/or time. For example, the immobilization of arsenic by microorganisms is an example of a symmetry-building behavior that leads to consistently low mobile soil arsenic levels both spatially and temporally. Klinger (2004) has shown that complex systems cannot attain long-term stability through symmetry-building behavior alone. Systems with only symmetry-building behavior become inflexible and unable to respond to change. The change could arise externally through a changing environment or internally through the natural evolution of a system. In the case of an ecosystem, there is a steady input of arsenic into the system through uptake by plants and weathering of sulfide minerals. Eventually, the added arsenic will overwhelm the ability of the microbial population to immobilize arsenic unless the ecosystem has some means of expelling arsenic from the system. It is proposed that the symmetry-breaking behavior that maintains the long-term stability is that the microbial population converts immobile arsenic back into mobile arsenic when leaching potential is high, so that excess arsenic can be leached from the system. This leads to the second preliminary hypothesis that mobile soil arsenic will increase at the beginning of a rainy season.

The second preliminary hypothesis can again be refined in light of the classification of soils as well as another characteristic of organisms, which is the ability to repel invasion by another organism. Emerman and Parmelee (2002) and Allen and Emerman (2003) showed data consistent with the hypothesis that a prairie ecosystem can resist invasion by a shrub ecosystem by maintaining low soil moisture and keeping nitrogen in organic form. If grasses are less susceptible to arsenic than woody plants due to the relatively rapid growth of grasses, a grassy ecosystem can resist invasion by woody plants by maintaining a sufficient level of soil arsenic. On that basis, it can be argued that the microbial population of grassy ecosystems should transfer arsenic into immobile form at the beginning of a rainy season in order to prevent the leaching of arsenic. However, when organic matter is low, the lack of microorganisms available for the sequestration of arsenic should cause even grassy ecosystems to transfer arsenic into mobile form at the beginning of a rainy season so that the microbial population does not become overwhelmed with arsenic.

In summary, the two preliminary hypotheses can be expanded into a set of five hypotheses as follows:

1) When mobile soil arsenic is compared across ecosystem boundaries in Entisols, mobile soil arsenic will be higher in the earlier stage of vegetational succession.
2) Mobile soil arsenic will not change across ecosystem boundaries in Mollisols.
3) In woody ecosystems, mobile soil arsenic will increase at the beginning of a rainy season.
4) In grassy ecosystems in Mollisols, mobile soil arsenic will decrease at the beginning of a rainy season.
5) In grassy ecosystems in Entisols, mobile soil arsenic will increase at the beginning of a rainy season.

For all of the above hypotheses, whether Inceptisols should be grouped with Entisols or Mollisols must be determined by experiment.

Testing the Hypotheses in Nepal

Emerman (2004) suggested that the Terai forest includes a microbial population that sequesters arsenic in immobile form. When the forest is cleared and replaced by agricultural land or intensively grazed pasture, the microbial population that sequesters arsenic can be lost, which could allow the mobilized arsenic to be leached into groundwater. Emerman (2004) measured mobile soil arsenic along traverses crossing sharp boundaries between forest and intensively grazed pasture at two sites in the Terai region. Both sites had soils in the order Inceptisols and measurements were made in the dry season (February) and the beginning of the monsoon season (June). Pasture, rather than agricultural land, is the appropriate comparison for forest since agricultural land in Bangladesh has been shown to contain high levels of soil arsenic due to irrigation with arsenic-contaminated groundwater (Islam et al. 2000; Meharg and Rahman 2003). At both sites mobile soil arsenic was higher on the pasture side than the forest side of the boundary in the dry season. At the beginning of the monsoon season, mobile soil arsenic rose on the forest side of the boundary at both sites so that mobile soil arsenic was the same on both sides of the boundary. The above results are consistent with the set of five hypotheses given above as long as Inceptisols are grouped with Entisols (soils with lower organic matter) in terms of understanding the spatial variation in mobile soil arsenic.

Testing the Hypotheses in Iowa

This section will concern further testing of the five hypotheses at three sites in Iowa, which correspond to the soil orders Entisols, Inceptisols and Mollisols. As these experiments have not previously been described in the scientific literature, they will be discussed in somewhat more detail than the experiments carried out in Nepal. A complete description of the experiments is available from the first author.

Description of study sites

The three soil orders of Entisols, Inceptisols and Mollisols were represented in Iowa by the three study sites at Behrens Ponds and Woodland State Preserve in Linn County, the Luhrs Property in Warren County (Fig. 5.1), and Rolling Thunder Prairie State Preserve in Warren County, respectively. These sites were chosen due to the presence of sharp boundaries between ecosystems. Within and to the east of Rolling Thunder Prairie, five ecosystems can be found within 300 m, which are row crops, pasture, prairie, deciduous shrubs and deciduous forest (Fig. 5.2). Since both the pasture and prairie are dominated by perennial grasses, they represent the same stage in the vegetational succession. Behrens Ponds and Woodland includes sand prairie, shrubs, savanna, wetlands and deciduous forest, but only the boundary between sand prairie and shrubs was examined in this study. At the Luhrs Property the boundary between prairie and shrubs was studied. At Rolling Thunder Prairie the forest occurs along the streams (Fig. 5.2). At all three sites shrubs are encroaching on prairie and the suppression of shrubs by fire, mowing and manual removal of woody plants are projects of the Linn County and Warren County Conservation Boards and the Luhrs family. The agricultural land to the east of Rolling Thunder Prairie was planted with corn prior to sampling (Fig. 5.2). The pasture east of Rolling Thunder Prairie is dominated by the perennial orchardgrass. The most abundant plants in the prairie are the grasses big bluestem and little bluestem. The dominant invading shrubs are roundleaf dogwood, smooth sumac and coralberry. Common trees along the streams include American elm and black willow (Mabry 2002). The sand prairie at Behrens Ponds and Woodland is dominated by the perennial grass smooth brome. Encroaching shrubs include black raspberry, American hazelnut, common pricklyash and seedlings of eastern cottonwood and black oak (Freese and Brown 2001). The prairie at the Luhrs Property is dominated by smooth brome with invasion by shrubs and vines false boneset, wild grape and multiflora rose.

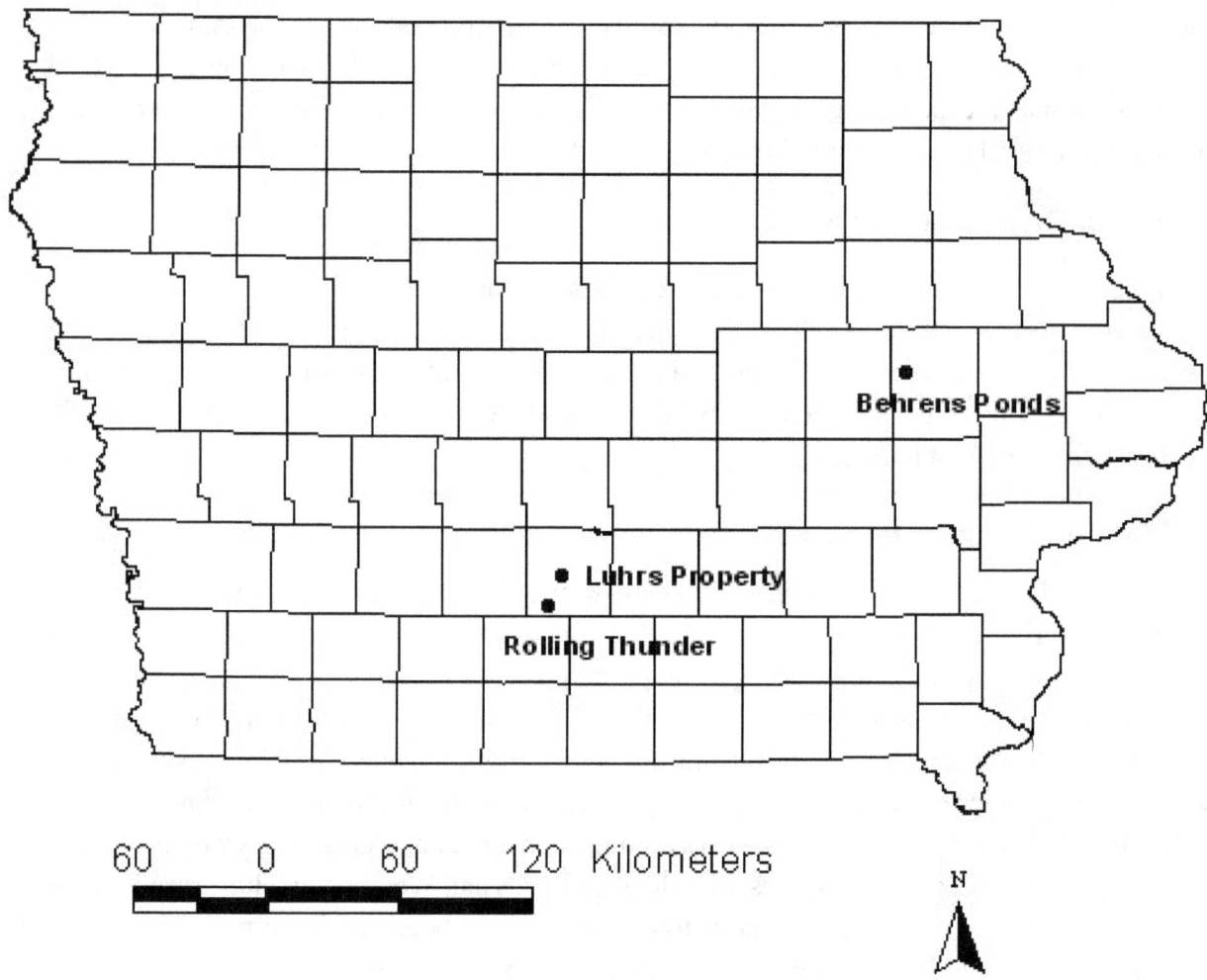

Fig. 5.1. Location of study sites superimposed on county map of Iowa. Mobile soil arsenic was measured in Mollisols at Rolling Thunder Prairie State Preserve in Warren County, in Inceptisols at the Luhrs Property in Warren County, and in Entisols at Behrens Ponds and Woodland State Preserve in Linn County.

Soil sampling and analysis

Soil samples were collected along five traverses crossing each ecosystem boundary, which included the row crop / pasture boundary, pasture / prairie boundary, prairie / shrub boundary and shrub / forest boundary at Rolling Thunder Prairie, and the prairie / shrub boundaries at the Luhrs Property and at Behrens Ponds and Woodland. Traverses were chosen where the ecosystem boundary appeared especially narrow and well-defined. At Rolling Thunder Prairie the ecosystems are arranged in order of increasing elevation as forest, shrubs, prairie, pasture, with row crops the same elevation as the pasture (Fig. 5.2). At the other two sites, shrubs are downslope from prairie. Therefore, samples were collected from four locations along each traverse corresponding to distances 5 m and 10 m on each side of the ecosystem boundary in

96

order to separate soil properties related to differences in topographic position from differences related to vegetation. Samples were collected from Rolling Thunder Prairie during the first two weeks of October 2003 and the first two weeks of April 2004, corresponding to the typical dry season and beginning of the rainy season in the Midwest. Samples were collected from the other two sites during the first two weeks of October 2004 and the first two weeks of April 2005. Sample locations were marked by flags and the spring samples were collected as close as possible to the fall samples without sampling the previously disturbed soil. Soil samples were removed corresponding to depths 0-5 cm, 25-30 cm and 35-40 cm at Rolling Thunder Prairie and to depths 5-10 cm and 25-30 cm at the Luhrs Property and at Behrens Ponds and Woodland. A total of 640 soil samples were collected and analyzed in this study. Mobile soil arsenic was measured by adding 50 mL of 1 M HCl to 5 g of field-moist soil (Langston 1980; Reuther 1992; Keon et al. 2001). The solution was stirred on a stir plate for 60 minutes and then poured directly into the wide-mouth reaction vessel of the Hach Arsenic Test Kit without filtering.

Results of experiment in Iowa

At Rolling Thunder Prairie (Mollisols site) there were no statistically significant differences in mobile soil arsenic across ecosystem boundaries for either fall or spring when the three depth measurements were averaged for each location (Fig. 5.3). When the depth measurements were considered separately, the only statistically significant difference in mobile soil arsenic occurred as higher mobile soil arsenic on the pasture side of the pasture / prairie boundary in the spring at depth 25-30 cm. On the other hand, the Rolling Thunder prairie site showed marked seasonal variation in mobile soil arsenic. When the three depth measurements were measured, mobile soil arsenic decreased for row crops at the row crop / pasture boundary, and for both pasture and prairie at the pasture / prairie boundary from fall to spring (Fig.5.3). Mobile soil arsenic did not change between fall and spring on the prairie side of the prairie / shrub boundary. When the three depth measurements were considered separately, mobile soil arsenic decreased in row crops at the row crop / pasture boundary from fall to spring at depths 0-5 cm and 25-30 cm. Mobile soil arsenic decreased in pasture only at depth 35-40 cm at both the row crop / pasture boundary and the pasture / prairie boundary. Mobile soil arsenic in prairie at the pasture / prairie boundary decreased from fall to spring only at depths 25-30 cm and 35-40 cm.

At Behrens Ponds and Woodland (Entisols site), there was no difference in mobile soil arsenic across the prairie / shrub boundary either when the two depths were averaged or considered separately (Fig. 5.4). At the Luhrs Property (Inceptisols site), mobile soil arsenic was greater on the prairie side of the boundary in the fall both when the depths were averaged and at the depth 25-30 cm. At Behrens Ponds and Woodland, when the two depths were averaged, mobile soil arsenic increased from fall to spring for both prairie and shrubs. In the prairie the change was statistically significant for both depths, while in the shrubs the change was statistically significant only at the depth 5-10 cm. At the Luhrs Property, mobile soil arsenic decreased from fall to spring in the prairie and did not change in the shrubs for both depths and for the average of the two depths. In no cases, at any of the sites, were differences between measurements 5 m and 10 m from an ecosystem boundary statistically significant. Therefore, measurements 5 m and 10 m from a boundary were averaged in all cases.

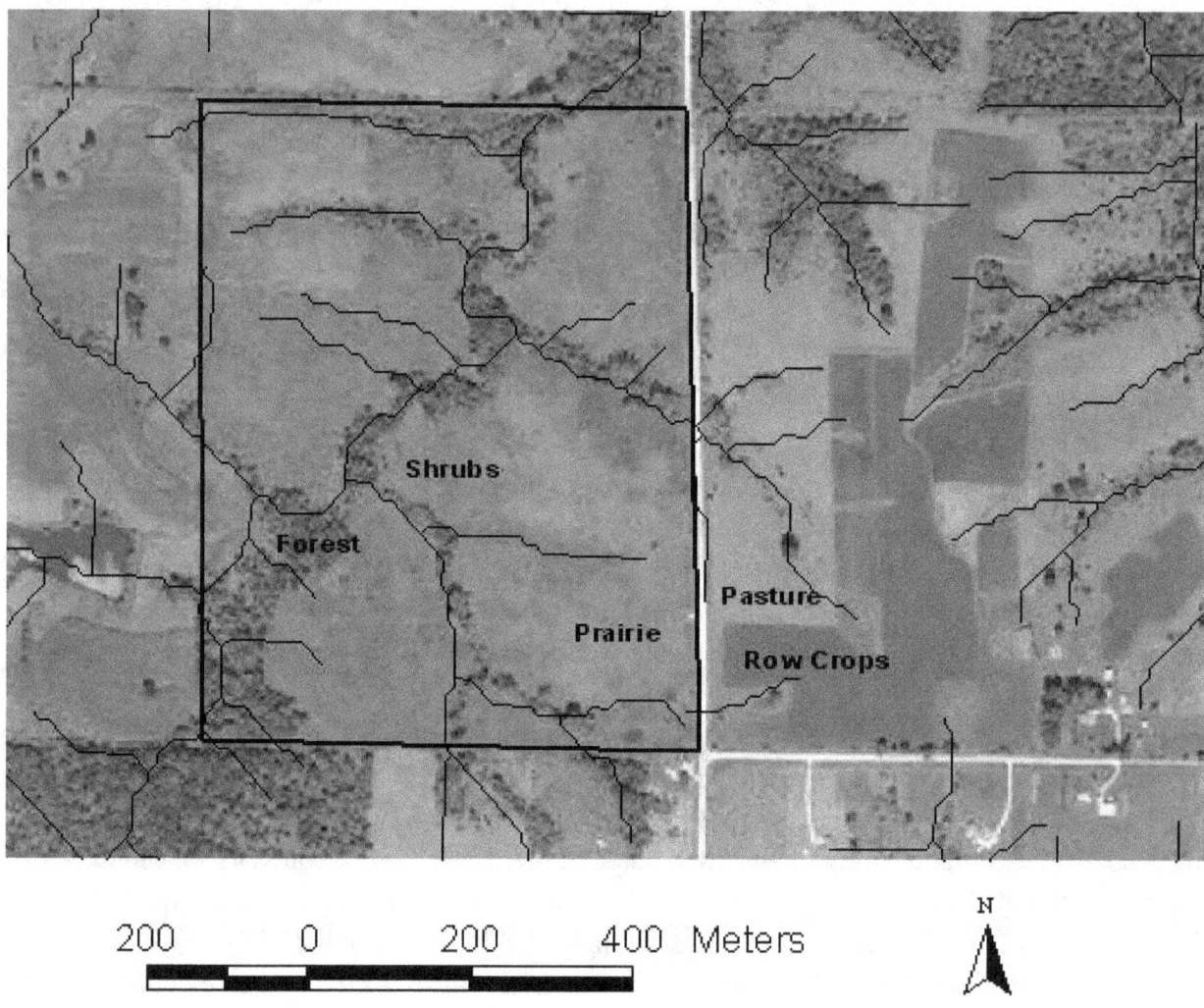

Fig. 5.2. Drainage map (ICSS and IGSB 1998) and outline of Rolling Thunder Prairie State Preserve superimposed on a portion of the aerial photo mosaic of Warren County (USDA 2004). Five ecosystems (row crops, pasture, prairie, shrubs and forest) can easily be identified on the

The results can be summarized as follows:

1) At the Mollisols and Entisols sites there were no changes in mobile soil arsenic across ecosystem boundaries. At the Inceptisols site mobile soil arsenic was higher on the prairie side of the prairie / shrub boundary.

98

2) At the Mollisols and Inceptisols sites mobile soil arsenic decreased from fall to spring for the ecosystems that did not include woody plants. At the Entisols site mobile soil arsenic increased from fall to spring for both prairie and shrubs.

The above results were not observed at all ecosystem boundaries for both seasons. There were ecosystem boundaries at which no change was observed when a change was predicted. However, there were no ecosystem boundaries at which an increase in mobile soil arsenic was observed when a decrease was predicted, or vice versa. The only significant prediction that was not satisfied was that there was no variation in mobile soil arsenic between prairie and shrubs at the Entisols site.

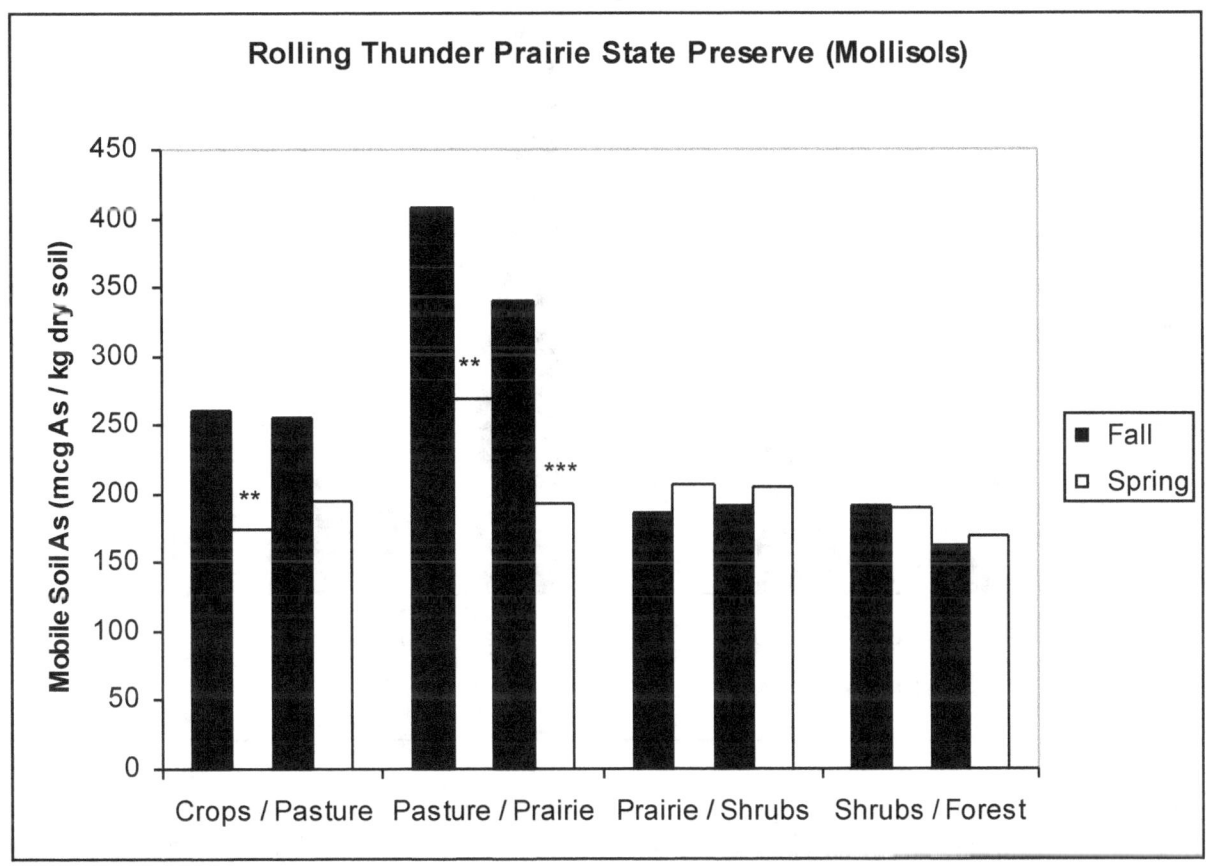

Fig. 5.3. Mobile soil arsenic concentration averaged over three depths (0-5 cm, 25-30 cm, 35-40 cm) on either side of four ecosystem boundaries at Rolling Thunder Prairie State Preserve (Mollisols site). The symbols *, *, *** indicate differences between fall and spring on a given side of an ecosystem boundary are statistically significant at the 95%, 99%, 99.9% confidence level according to the Student's unpaired t-test. No differences across ecosystem boundaries for a given season are statistically significant at the 95% confidence level.

The results from the three sites in Iowa are broadly consistent with the set of five hypotheses as long as (1) Inceptisols are grouped with Entisols in terms of understanding the spatial variation in mobile soil arsenic, and (2) Inceptisols are grouped with Mollisols in terms of understanding the temporal variation in mobile soil arsenic. The above dichotomy is consistent with results from Nepal (Emerman 2004). There does not seem to be any reason why the same critical level of soil organic matter must be applicable to understanding both temporal and spatial variation.

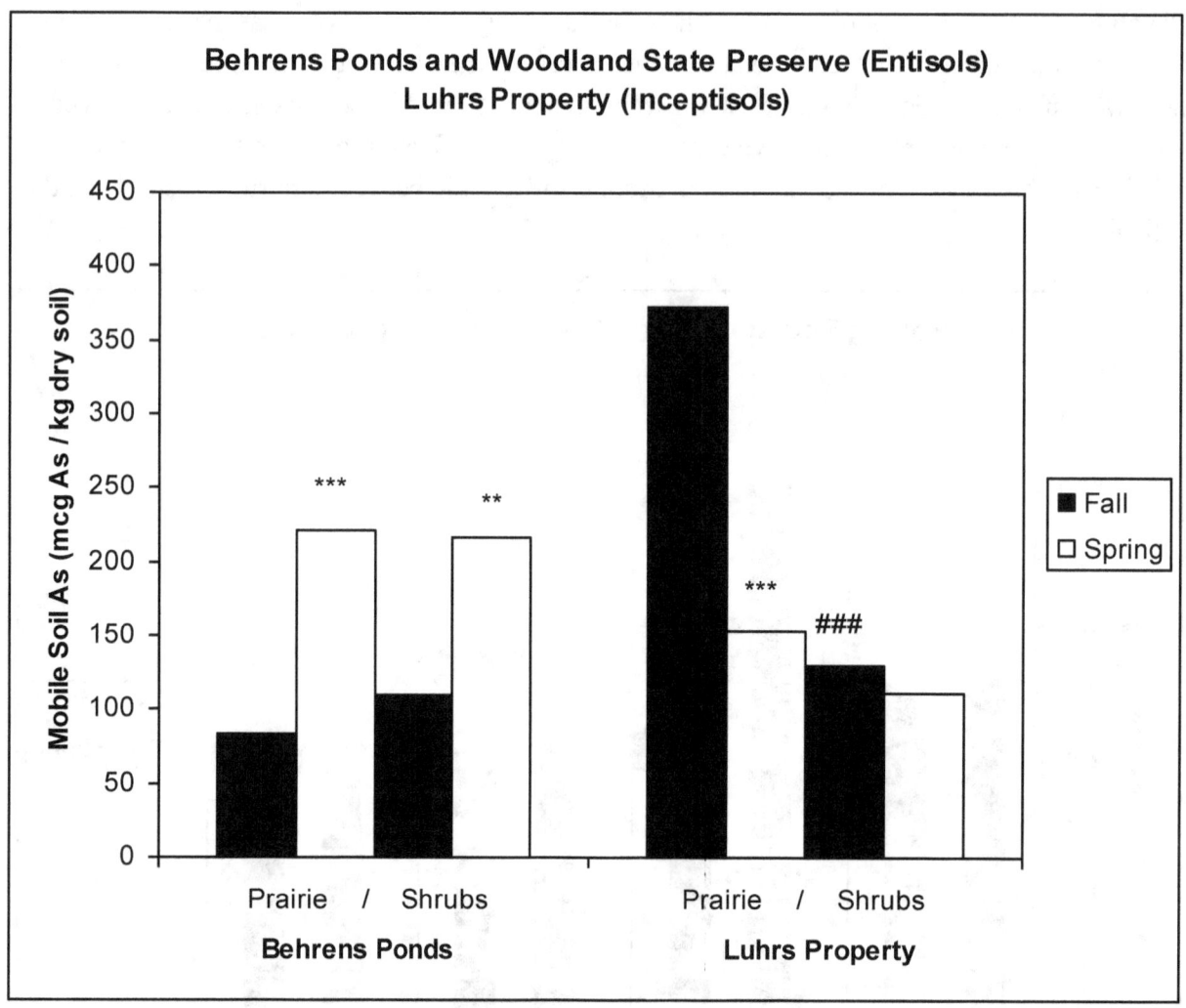

Fig. 5.4. Mobile soil arsenic concentration averaged over two depths (5-10 cm, 25-30 cm) on either side of prairie / shrub boundaries at Behrens Ponds and Woodland State Preserve (Entisols site) and the Luhrs Property (Inceptisols site). The symbols *, *, *** indicate differences between fall and spring on a given side of an ecosystem boundary are statistically significant at the 95%, 99%, 99.9% confidence level according to the Student's unpaired t-test.

Recommendations for Addressing Arsenic Contamination of Groundwater in Nepal

Conservation and restoration of pre-existing vegetation

We are proposing that arsenic contamination of groundwater in the Terai region of Nepal does result from human activity and results from the dramatic changes in land use that have occurred since the settlement of the Terai region in the 1950s. It is not simply deforestation, but also a loss of healthy grasslands, that causes arsenic contamination of groundwater. The preliminary data collected in Nepal showed that mobile soil arsenic was greater in intensively grazed pasture than forest only during the dry season. During the monsoon season, when leaching could occur, the forest mobile soil arsenic rose to the level of the pasture mobile soil arsenic (Emerman 2004). On that basis, it could be argued that deforestation is irrelevant to groundwater contamination. The key point is that the measurements in Nepal were made in Inceptisols. This study has shown that, in Inceptisols and Mollisols, while mobile soil arsenic will increase in forest during the rainy season, mobile soil arsenic will increase in grasslands during the dry season. (The intensively grazed pastures studied by Emerman (2004) cannot be regarded as healthy grasslands.) On the other hand, in Entisols, mobile soil arsenic will increase in grasslands during the rainy season. Leaching of arsenic will be minimized whenever mobile soil arsenic is low during the rainy season. Therefore, the vegetation cover that would minimize leaching of arsenic would be grasslands on the Inceptisols and Mollisols. From the standpoint of minimizing leaching of arsenic, conservation is less important on the Entisols. But if any vegetation cover needs to be conserved on the Entisols, it is forest rather than grassland. In the Terai region of Nepal, Entisols are found along the active alluvial plains (river channels, sand and gravel bars, and low terraces) with Inceptisols and Mollisols everywhere else (Nepal Survey Department 1982).

We are suggesting that further arsenic contamination of groundwater in the Terai region of Nepal could be prevented by conserving forest along the rivers and grasslands everywhere else. The Siwalik region of Nepal (foothills of the Himalaya) also includes Entisols along the rivers with Inceptisols, Mollisols and some Alfisols (soils that develop under deciduous forest) everywhere else, except for the steep slopes and mountainous terrain that are underlain by Entisols (Nepal Survey Department 1982). By the same reasoning, in the Siwalik region, forests should be conserved on the rivers and steep slopes with grasslands conserved everywhere else. Soils have not been mapped in the rest of Nepal (Lesser Himalayan and Higher Himalayan zones), but the steep slopes and mountainous terrain must correspond with the dominance of Entisols. On that basis, there should be as much conservation of forest as possible in the Lesser and Higher Himalayan zones. We are suggesting that the above description of a vegetation cover that results in maximum immobilization of arsenic in soil corresponds to Leopold's (1949) concept of healthy land maintaining itself as an organism.

Soil management based on the Gaia Hypothesis

The conservation and restoration of pre-existing vegetation can easily be viewed as a "Return to the Garden of Eden" plan even if the plan involves only setting priorities as to which landscapes and ecosystems most urgently require conservation and restoration. There could be many social reasons as to why deforestation and destruction of grasslands cannot be reversed, including the needs of an expanding population (who have just as much right to food and shelter as everyone else). The implication of this study is that the immobilization of arsenic in soil results not simply from a particular vegetation cover on a particular soil order, but from the microbial community that would normally be associated with a particular combination of vegetation cover and soil order. Therefore, the question arises as to whether the microbial community could be restored without the full restoration of the vegetation cover.

As discussed above, forests are in decline on a worldwide basis as increased environmental acidity causes a shift to later stages of vegetational succession. Lee Klinger's institute Sudden Oak Life has demonstrated that the vegetational shift can be reversed, i.e., the attack of trees by mosses can be halted, by applying an alkaline-rich calcium carbonate paste to trees, by applying alkaline-rich volcanic ash to soils, or by periodic fires, which also promote soil alkalinity (Sudden Oak Life 2011). In the case of the Terai region of Nepal, the destruction of healthy forests and grasslands is equivalent to shifting the vegetational succession to the earlier stages, to somewhere between bare soil and annual grasses. Therefore, it is possible that the microbial population of a healthy forest or grassland could be restored by an appropriate intentional increase in soil acidity, which possibly could be accomplished by appropriate agricultural and pastoral practices. The important point is that an investigation into the implications of regarding ecosystems as organisms has led to new possibilities for solutions to an environmental problem that disproportionately affects the poor.

Acknowledgements

We thank the Linn County and Warren County Conservation Boards for permission to collect soil samples. This research was partially funded by an undergraduate research grant from the Beta Beta Beta Research Scholarship Foundation.

References

Acharyya, S.K., P. Chakraborty, S. Lahiri, B. C. Raymahashay, S. Guha, and A. Bhowmik, A., 1999. Arsenic poisoning in the Ganges delta. Nature, v. 401, p. 545.

Acharyya, S. K., S. Lahiri, B. C. Raymahashay, and A. Bhowmik, 2000. Arsenic toxicity of groundwater in parts of the Bengal basin in India and Bangladesh: The role of Quaternary stratigraphy and Holocene sea-level fluctuation. Environmental Geology, v. 39, pp. 1127-1137.

Allen, N. S. and S. H. Emerman, 2003. The control of the nitrogen cycle as a mechanism for the self-organization of prairie ecosystems. *In* Proceedings of the 18th North American Prairie Conference, S. Foré (ed.). Truman State University Press, Kirksville, MO, pp. 151-156.

Anderson, R. C., 2006. Evolution and origin of the Central Grassland of North America: Climate, fire and mammalian grazers. Journal of the Torrey Botanical Society, v. 133, pp. 626-647.

Appelo, M., M. J. J. Van der Weiden, C. Tournassat, and L. Charlet, 2002. Surface complexation of ferrous iron and carbonate on ferrihydrite and the mobilization of arsenic. Environmental Science and Technology, v. 36, pp. 3096-3103.

Aswathanarayana, U., 1997. Arsenic in groundwater, West Bengal. Journal of the Geological Society of India, v. 49, pp. 341-345.

Badal, K. M., T. Roy Choudhury, G. Samanta, G. K. Basu, P. P. Chowdhury, C. R. Chanda, D. Lodh, N. K. Karan, R. K. Dhar, D. K. Tamili, D. Das, K. C. Saha, and D. Chakraborti, 1996. Arsenic in groundwater in seven districts of West Bengal, India—The biggest As calamity in the world. Current Science, v. 70, pp. 976-985.

Bak, P., 1999. How Nature Works: The Science of Self-Organized Criticality. Springer-Verlag, New York.

Bar-Yam, Y., 2003. Dynamics of Complex Systems (Studies in Nonlinearity). Westview, Boulder, CO.

Bhattacharaya, P., D. Chatterjee, and G. Jacks, 1997. Occurrence of As-contaminated groundwater in alluvial aquifers from the Delta Plains, Eastern India: Options for safe drinking water supply. Water Resources Development, v. 13, pp. 79-92.

Bhattacharya, P., N. Tandukar, A. Neku, A. A. Valero, A. B. Mukherjee, and G. Jacks, 2003. Geogenic arsenic in groundwaters from Terai Alluvial Plain of Nepal. Journal de Physique IV (France), v. 107, pp. 173-176.

Bose, P. and A. Sharma, 2002. Role of iron in controlling speciation and mobilization of arsenic in subsurface environment. Water Resources, v. 36, pp. 4916-4926.

Bradbury, R. H., D. G. Green, and N. Snoad, 2000. Are ecosystems complex systems? *In* Complex Systems, T. R. J. Bossomaier and D. G. Green (eds.). Cambridge University Press, Cambridge, UK, pp. 339-365.

Brady, N. C. and R. R. Weil, 2007. The Nature and Properties of Soils, 14th ed. Prentice Hall, Upper Saddle River, NJ.

Clements, F. E., 1916. Plant Succession: An Analysis of the Development of Vegetation. Publication 242, Carnegie Institution of Washington, Washington, D.C.

Coveney, P. and R. Highfield, 1996. Frontiers of Complexity: The Search for Order in a Chaotic World. Faber and Faber, London.

Cullen, W. R. and K. J. Reimer, 1989. Arsenic speciation in the environment. Chemical Reviews, v. 89, pp. 713-764.

Dhar, R. K., B. K. Biswas, G. Samanta, B. K. Mandal, D. Chakraborti, S. Roy, A. Jafar, A. Islam, G. Ara, S. Kabir, A. W. Khan, S. K. Ahmed, S. A. Hadi, 1997. Groundwater As calamity in Bangladesh. Current Science, v. 73, pp. 48-59.

Downing, K. and P. Zvirinsky, 1999. The simulated evolution of biochemical guilds: Reconciling Gaia theory and natural selection. Artificial Life, v. 5, pp. 291-319.

Emerman, S.H., 2004. Deforestation, arsenic, and the self-organizing jungle in the Terai region of Nepal. Journal of Nepal Geological Society, v. 29, pp. 13-22.

Emerman, S.H., 2005. Arsenic and other heavy metals in the rivers of central Nepal. Journal of Nepal Geological Society, v. 31, pp. 11-18.

Emerman, S. H, and Parmelee, J. R., 2002. The control of infiltration as a mechanism for the self-regulation of prairie ecosystems: Preliminary studies at Rolling Thunder Prairie State Preserve, Warren County, Iowa. *In* Proceedings of the 22nd Annual American Geophysical Union Hydrology Days, J. A. Ramirez (ed.). Hydrology Days Publications, Atherton, CA, pp. 76-85.

Emerman, S. H., T. N. Bhattarai, D. P. Adhikari, S. R. Joshi, S. L. Lakhe, A. J. Luhrs, K. R. Prasai, and K. L. Robson, 2007. Origin of arsenic and other heavy metals in the rivers of Nepal. Journal of Nepal Geological Society, v. 35, pp. 29-36.

Emerman, S. H., T. Prasai, R. B. Anderson, and M. A. Palmer, 2010. Arsenic contamination of groundwater in the Kathmandu Valley, Nepal, as a consequence of rapid erosion. Journal of Nepal Geological Society, v. 40, pp. 49-60.

Emerman, S. H., R. B. Anderson, S. Bhandari, R. R. Bhattarai, M. A. Palmer, T. N. Bhattarai, and M. P. Bunds, 2011. Arsenic and other heavy metals in the Sunkhoshi and Saptakhoshi Rivers, eastern Nepal. Journal of Nepal Geological Society, v. 43, pp. 241-254.

ENPHO (Environment and Public Health Organization) and USGS (United States Geological Survey), 2004. The state of arsenic 2003 in Nepal (a draft report). National Arsenic Steering Committee (NASC), Kathmandu, Nepal.

Freese, E. L. and M. E. Brown, 2001. Behrens Ponds and Woodland State Preserve, Linn County, Iowa, Vascular Plant List. Unpublished report to the State Preserves Advisory Board (Iowa).

Gustafsson, J. P. and G. Jacks, 1995. Arsenic geochemistry in forested soil profiles as revealed by solid-phase studies. Applied Geochemistry, v. 10, pp. 307-315.

Harding, S.P., 1999. Food web complexity enhances community stability and climate regulation in a geophysiological model. Tellus, v. 51B, pp. 815-829.

Harvey, C. F., C. H. Swartz, A. B. M. Badruzzaman, N. Keon-Blute, W. Yu, M. A. Ali, J. Jay, R. Beckie, V. Niedan, D. Brabander, P. M. Oates, K. N. Ashfaque, S. Islam, H. Hemond, and M. F. Ahmed, 2002. Arsenic mobility and groundwater extraction in Bangladesh. Science, v. 298, pp. 1602-1606.

Holland, J., 1996. Hidden Order: How Adaptation Builds Complexity. Basic Books, New York.

Islam, M. R., P. Lahermo, R. Salminen, S. Rojstaczer, and V. Peuraniemi, 2000. Lake and reservoir water quality affected by metals leaching from tropical soils, Bangladesh. Environmental Geology, v. 39, pp. 1083-1089.

ICCS (Iowa Cooperative Soil Survey) and IGSB (Iowa Geological Survey Bureau), 1998. DRAIN_91 (vector digital data). Iowa Geological Survey Bureau, Iowa City, IA. Available online at http://www.igsb.uiowa.edu/nrgislibx/.

Jørgensen, S. E., B. C. Patten, and M. Straskraba, 1992. Ecosystems emerging: Toward an ecology of complex systems in a complex future. Ecological Modeling, v. 62, pp. 1-27.

Jørgensen, S. E. and F. Müller, 2000. Ecosystems as complex systems. *In* Handbook of Ecosystem Theories and Management, S. E. Jørgensen (ed.). Lewis Publications, New York, pp. 5-20.

Kanel, S. R., H. Choi, K. W. Kim, and S. H. Moon, 2005. Arsenic contamination in groundwater in Nepal: a new perspective and more health threat in South Asia. *In* Natural Arsenic in Groundwater: Occurrence, Remediation and Management, J. Bundschuh, P. Bhattacharya and D. Chandrasekharam (eds.). A. A. Balkema Publishers, Leiden, The Netherlands, pp. 103-108.

Kauffman, S. A., 1993. The Origins of Order: Self-Organization and Selection in Evolution. Oxford University Press, New York.

Kauffman, S. A., 2000. Investigations. Oxford University Press, New York.

Keon, N. E., C. H. Swartz, D. J. Brabander, C. Harvey, and H. F. Hemond, 2001. Validation of an arsenic sequential extraction method for evaluating mobility in sediments. Environmental Science and Technology, v. 35, pp. 2778-2784.

Klinger, L. F., 1990. Global patterns in community succession, 1. Bryophytes and forest decline. Memoirs of the Torrey Botanical Club, v. 24, pp. 1-50.

Klinger, L. F., 1991. Peatland formation and ice ages: A possible Gaian mechanism related to community succession. *In* Scientists on Gaia, S. H. Schneider and P. J. Boston (eds.). The MIT Press, Cambridge, MA, pp. 246-255.

Klinger, L. F., 1996a. The myth of the classic hydrosere model of bog succession. Arctic and Alpine Research, v. 28, pp. 1-9.

Klinger, L. F., 1996b. Coupling of soils and vegetation in peatland succession. Arctic and Alpine Research, v. 28, pp. 380-387.

Klinger, L. F., 2004. Gaia and complexity. *In* Scientists Debate Gaia: The Next Century, S. H. Schneider, J. R. Miller, E. Crist, and P. J. Boston (eds.). The MIT Press, Cambridge, MA, pp. 187-200.

Klinger, L. F., S. A. Elias, V. M. Behan-Pelletier, and N. E. Williams, 1990. The bog climax hypothesis: Fossil arthropod and stratigraphic evidence in peat sections from southeast Alaska, USA. Holarctic Ecology, v. 13, pp. 72-80.

Klinger, L. F., P. R. Zimmerman, J. P. Greenburg, L. E. Heidt, and A. B. Guenther, 1994. Carbon trace gas fluxes along a successional gradient in the Hudson Bay lowland. Journal of Geophysical Research, v. 99, pp. 1469-1494.

Klinger, L. F. and S. K. Short, 1996. Succession in the Hudson Bay lowland, northern Ontario, Canada. Arctic and Alpine Research, v. 28, pp. 172-183.

Klinger, L. F., J. A. Taylor, and L. G. Franzen, 1996. The potential role of peatland dynamics in ice-age initiation. Quaternary Research, v. 45, pp. 89-92.

Klinger, L. F. and D. J. Erickson III, 1997. Geophysiological coupling of marine and terrestrial ecosystems. Journal of Geophysical Research, v. 102, pp. 25,359-25,370.

Klinger, L. F., J. Greenburg, A. Guenther, G. Tyndall, P. Zimmerman, M. M'Bangui, J.-M. Moutsamboté, and D. Kenfack, 1998. Patterns in volatile organic compound emissions along a savanna-rainforest gradient in central Africa. Journal of Geophysical Research, v. 103, pp. 1443-1454.

Langston, W. J., 1980. Arsenic in the U.K. estuarine sediments and its availability to benthic organisms. Journal of the Marine Biology Association, United Kingdom, v. 60, pp. 869-881.

Lenton, T. M. and M. van Oijen, 2002. Gaia as a complex adaptive system. Philosophical Transactions of the Royal Society (Series B), v. 357, pp. 683-695.

Leopold, A., 1949. A Sand County Almanac, and Sketches Here and There. Oxford University Press, New York.

Levin, S. A., 1998. Ecosystems and the biosphere as complex adaptive systems. Ecosystems, v. 1, pp. 431-436.

Lewin, R., 2000. Complexity: Life at the Edge of Chaos. University of Chicago Press, Chicago, IL.

Lovelock, J., 1995. The Ages of Gaia: A Biography of our Living Earth. W.W. Norton, New York.

Mabry, C., 2002. Floristic inventory of Rolling Thunder Prairie State Preserve. Unpublished report to the State Preserves Advisory Board (Iowa).

Mallick, S. and N. R. Rajgopal, 1996. Groundwater development in the arsenic-affected alluvial belt of West Bengal—some questions. Current Science, v. 70, pp. 956-958.

Margalef, R., 1963. On certain unifying principles of ecology. American Naturalist, v. 97, pp. 357-375.

McArthur, J. M., P. Ravenscroft, S. Safiullah, and M. F. Thirlwall, 2001. Arsenic in groundwater: Testing pollution mechanisms for sedimentary aquifers in Bangladesh. Water Resources Research, v. 37, pp. 109-117.

Meharg, A. A. and M. Rahman, 2003. Arsenic contamination of Bangladesh paddy field soils: Implications for rice contribution to arsenic consumption. Environmental Science and Technology, v. 37, pp. 229-234.

Nepal Survey Department, 1982. Land Systems Map, Rupandehi, Nawalparasi and Palpa Districts 1:50,000, Nepal Western Development Region (Sheet No. 63 M/10). Topographical Survey Branch, Department of Survey, Ministry of Land Reform, Kathmandu, Nepal.

Nickson, R. T., J. M. McArthur, W. G. Burgess, K. M. Ahmed, P. Ravenscroft, and M. Rahman, 1998. Arsenic poisoning of Bangladesh groundwater. Nature, v. 395, p. 338.

Nickson, R. T., J. M. McArthur, P. Ravenscroft, W. G. Burgess, and K. M. Ahmed, 2000. Mechanism of arsenic release to groundwater, Bangladesh and West Bengal. Applied Geochemistry, v. 15, pp. 403-413.

Odum, E. P., 1969. The strategy of ecosystem development. Science, v. 164, pp. 262-270.

Rayner, A. D. M., 1997. Degrees of Freedom: Living in Dynamic Boundaries. Imperial College Press, London.

Reuther, R., 1992. Arsenic introduced into a littoral freshwater model ecosystem. Science of the Total Environment, v. 115, pp. 219-237.

Shanker, R., T. Pal, P. K. Mukherjee, S. Shome, and S. Sengupta, 2001. Association of microbes with arsenic-bearing siderite concretions from shallow aquifer sediments of Bengal delta and its implication. Journal of the Geological Society of India, v. 58, pp. 269-271.

Sharma, C.K., 1988. Natural Hazards and Man-Made Impacts in the Nepal Himalaya. Mrs. Pushpa Sharma, Kathmandu, Nepal.

Sharma, C.K., 1995. Some Symptoms of Environmental Degradation in Nepal (1950-1994). Mrs. Sangeeta Sharma, Kathmandu, Nepal.

Smedley, P. L. and D. G. Kinniburgh, 2002. A review of the source, behavior and distribution of arsenic in natural waters. Applied Geochemistry, v. 17, pp. 517-568.

Soil Survey Staff, 2007. Keys to Soil Taxonomy. Pocahontas Press, Blacksburg, VA.

Solé, R. V. and S. C. Manrubia, 1995. Are rainforests self-organized in a critical state? Journal of Theoretical Biology, v. 173, pp. 31-40.

Sudden Oak Life, 2011. Sudden Oak Life: Observations on oak health, tree care, organic farming, gardening, forest decline, acid rain, climate change, Gaia… by Lee Klinger. Available online at http://suddenoaklifeorg.wordpress.com/.

Tandukar, N., P. Bhattacharya, G. Jacks, and A. A. Valero, 2005. Naturally occurring arsenic in groundwater of Terai region in Nepal and mitigation options. In Natural Arsenic in Groundwater: Occurrence, Remediation and Management, J. Bundschuh, P. Bhattacharya, and D. Chandrasekharam (eds.), A.A. Balkema Publishers, Leiden, The Netherlands, pp. 41-48.

USDA (United States Department of Agriculture), 2004. National Agriculture Imagery Program (NAIP) aerial photography for Warren County, Iowa. United States Department of Agriculture, FSA Aerial Photography Field Office, Salt Lake City, Utah. Available online at http://www.igsb.uiowa.edu/nrgislibx/.

Von Bloh, W., A. Block, and H. J. Schellnhuber, 1997. Self-stabilization of the biosphere under global change: a tutorial geophysiological approach. Tellus, v. 49B, pp. 249-262.

Watson, A. J. and J. E. Lovelock, 1983. Biological homeostasis of the global environment: The parable of Daisyworld. Tellus, v. 35B, pp. 284-289.

Part II: Recruiting Citizen Scientists: Inclusion and Exclusion in Community-Level Environmental Activism

Chapter 6: Public Participation and Spatial Information

Shalini P. Vajjhala

Engaging citizens and communities is increasingly part of both science and policy agendas. From local level development planning to global environmental management, public participation is inherently based on spatial information. The locations of communities and resources are critical to a wide variety of decisions. As a result, various mapping methodologies—from hand drawn participatory map-making to detailed geographic information systems (GIS) analysis—have become particularly important for evaluating individual and community priorities, perceptions, and preferences for their changing environments. Recent developments in mapping tools and methods have transformed participation, affecting who participates, when, how, and how often. Although these efforts are directed at making participatory processes more inclusive and effective, in some cases massive quantities and highly sophisticated presentations of data have resulted in a divide beyond a lack of access to technology or information. This new divide—between information and communication—is evident where various stakeholders and diverse groups require common information about a project, but understand and use this information very differently from one another. In some cases, information is both available and relevant, but it is presented in a form that is too general or too specific to be useful for the intended audience. This chapter highlights the challenges and opportunities of facilitating participation at the intersection of science and social justice, where mapping not only has the potential to create "seats at the table" for underrepresented and underserved populations, but also has the potential to expand the table altogether.

Introduction

The Danish philosopher Søren Kierkegaard once stated, "If one is truly to succeed in leading a person to a specific place, one must first and foremost take care to find him where he is and begin there" (Kierkegaard et al. 1998). This observation is at the core of long-standing efforts to improve public participation. Engaging citizens and communities is increasingly part of both science and policy agendas, affecting a wide range of issues from local development planning to global environmental management. There is growing recognition in scientific and policy communities that improving participation across scales, groups, and issues is critical to building long-term support for policy implementation. Yet, doing so requires new approaches to information exchange, dialogue, and decision making.

Nowhere is this need more evident than in the tug-of-war between contrasting development and environmental agendas around the world. Take for example the issue of global climate change. The emerging challenges posed by climate change bring into sharper focus the need for public participation at ever larger scales and greater levels of local engagement. On the one hand, characterizing potential climate impacts requires both more detailed and more comprehensive scientific data on dynamic natural systems and their potential relationship with human activities and impacts. Gathering this information depends on understanding how environmental changes are likely to affect various populations with different patterns of use and reliance on natural resources. On the other hand, managing these challenges demands unprecedented levels of policy coordination, where implementing any new initiatives hinges on broad public acceptance.

In contrast, many recent examples of public participation have become synonymous with public opposition. From finding locations for new energy facilities to conserving natural resources, traditional approaches to siting and planning, often known as decide-announce-defend processes, have frequently failed. These processes have since become efforts that are decided-announced-defended... and abandoned (Kasperson et al. 1980). A variety of terms associated with this transition have entered the public vocabulary. From the familiar NIMBY, not-in-my-backyard, to the more recent BANANA, build-absolutely-nothing-anywhere-near-anything, these acronyms hint at the magnitude of the problems associated with encouraging more inclusive, effective public participation in science and policy (Inhaber 1998).

Equity and justice are at the crux of both of these debates. Developing new scientific research and designing policy instruments that take into account impacts on minority, indigenous, and poor communities are essential to ensuring socially and environmentally sensitive, sustainable decisions. So as scientists and citizens, what are our options for moving forward? This chapter highlights the changing roles of public participation in science and policy, and discusses the potential for mapping and spatial information to serve as a bridge between scientists, citizens, and decision makers.

"Unpacking" Participation

Public participation is a long-standing phenomenon evoking images ranging from the times of Roman forums to modern town-hall meetings. Despite the many, broad uses of the term, participatory projects and studies are often highly context specific, focusing on particular methods, tools, or places. As a result, the vast literature and practice associated with the concept of participation can be broadly placed into five categories: definitions, goals, methods, applications, and outcomes. Given the depth and breadth of participatory work, a few key examples from each of these five areas are highlighted here to provide background and context for the links between science, policy, and social justice.

What is participation?

The most general area of research on participation focuses on defining the experience itself. Studies in this area are largely based in the field of planning, and center on characterizing degrees of public inclusion and evaluating different levels of stakeholder involvement. Initial critical evaluations of citizen participation in the United States emerged in the wake of urban renewal programs, public health and welfare projects, and public administration efforts in the late 1950s and 1960s. These studies focused on bringing structure to the vague, top-down notions of participation of the time.

Two of the earliest and most influential evaluations in this area are Edmund Burke's (1968) Citizen Participation Strategies and Sherry Arnstein's (1969) A Ladder of Citizen Participation. Prior to these studies, all participation (in both theory and practice) was viewed positively as a general effort to engage citizens and public stakeholders. Arnstein's ladder dispels this notion and establishes eight "rungs" of public involvement that range from levels of non-participation, such as manipulation and persuasion, to levels of citizen power, such as partnership and civic control (Fig. 6.1). These original value-based characterizations of participation have since been widely adapted to changing attitudes about and approaches to participation.

Building Blocks of Participation

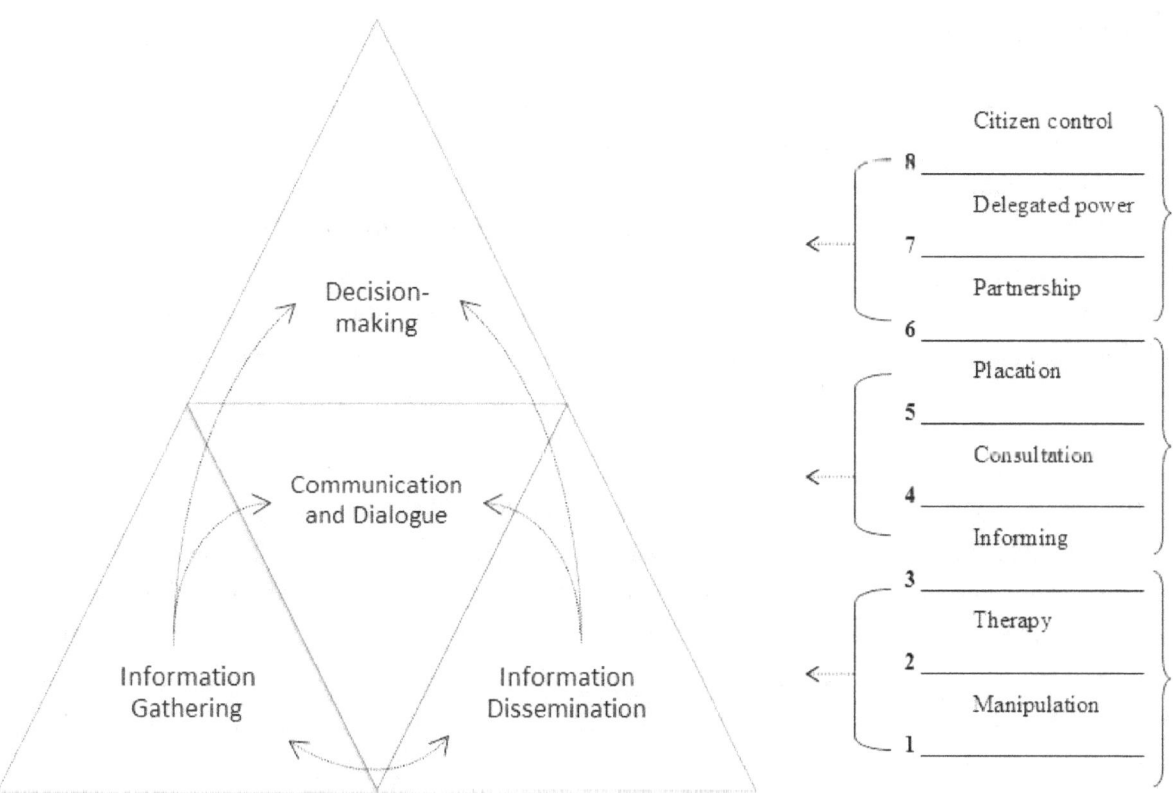

Fig. 6.1. Building blocks of citizen participation (left) and Arnstein's "Ladder of Citizen Participation" (right).

Participation today is rarely described in as general terms as it once was, and participatory studies now differentiate among public, citizen, and stakeholder involvement, and various forms of voluntary, solicited, and strategic participation, among others (Cornwall and Jewkes 1995; Renn et al. 1995; World Bank 1995; Webler 1999; Rowe and Frewer 2000; Irvin and Stansbury

2004). Since the early evaluations by Arnstein (1969) and Burke (1968), it has also become commonplace for agencies and organizations around the world to promote participatory processes and develop their own relevant definitions of participation. For example, the U.S. Environmental Protection Agency (EPA) has over the years supported various working groups on stakeholder involvement in environmental decision making and published numerous reports defining and describing different types of participation (EPA 2005). These adaptations have extended participation-related work to the fields of democracy, civil rights, consensus building, social movements, risk communication, public policy, and environmental justice, among others (Fiorino 1990; Chambers 1994; Pretty 1995; Davis and Whittington 1998; Chess and Purcell 1999; McDaniels et al. 1999; Beierle and Cayford 2002; Arvai 2003; Kurtz 2003, 2007). However, with the increasing specificity of participatory projects, even general works on participation have become more specialized, bringing us to the second broad area of participatory research and practice.

Why is participation important?

Just as efforts to define participation are wide-ranging, efforts to characterize the goals of participatory processes are also extremely broad. There are many motivations for choosing to support participation at all, either as an organizer or as a contributor. Some common reasons for engaging in participatory processes include: educating or informing the public, filling in gaps between local and expert knowledge, incorporating public values into decisions, improving the quality and acceptability of decisions, building trust among stakeholders and institutions, and resolving conflict among competing interests. Even among those who share these goals, there is tremendous diversity in how objectives can be met.

Depending on participants' priorities, the focus of participatory processes can range from short- to long-term, individual to societal, and process- to outcome-oriented agendas, to name a few. Among these there are hosts of related short-term, project-specific, and outcome-based objectives. These could include gathering local perceptions of proposed development plans, assessing variations in exposure to a specific environmental hazard, communicating risk, and negotiating over the siting of a major energy facility. In this vein, the National Research Council (1996) defines one of the primary goals of participation as improving the acceptability of risk and policy decisions.

In many cases, just as characterizations of participation have grown increasingly context specific, the aims and intentions of participation have also become equally specialized. Given the complexities and uncertainties surrounding the most basic goals and objectives of participation, even the most well-intentioned and organized of participatory processes could fail. As a result, the particular tools and methods for facilitating participation have drawn greater attention in an effort to reduce uncertainty and improve both public engagement efforts and outcomes.

How is participation implemented?

There are a variety of ways to initiate and sustain participatory processes. A few common methods of public engagement include forums, citizen advisory panels, consensus conferences, town meetings, household surveys, citizens' juries, and related methods. Agencies have developed manuals and toolkits for participation in different contexts. As participatory strategies have become highly tailored to different applications, the focus has shifted from methods to tools for improving the depth and breadth of outreach efforts.

In recent years, the most widespread tools for supporting stakeholder involvement have come from the information and communications technology sector. Internet-based dialogues have transformed both solicited and voluntary participation efforts. Furthermore, global initiatives have made technology a cornerstone of participatory development planning and environmental decision making. With the rapid spread of technologies from internet-based information dissemination and communication to spatial data infrastructures, opportunities for engaging stakeholders and the general public have entered a variety of previously top-down projects and fields.

As a result, critical evaluations of participatory methodologies have also become prominent. Aspects of current research on participation include broader discussions on "appropriate technology" and "the digital divide," focusing directly on the implications of using tools for facilitating participation in settings where a technology could drive a social process. In response to these criticisms, various tools have become progressively more sophisticated, focusing broadly on how participation is facilitated on the ground and how current participatory strategies could be improved (Chambers 1983; Yapa 1991; Dunn et al. 1997; Abbot et al. 1998; Brodnig and Mayer-Schönberger 2000; Cernea and McDowell 2000; Craig et al. 2002; McCall 2003; Tripathi and Bhattarya 2004; Vajjhala 2006).

Where and how are participatory tools and methods used?

Conventionally, participation as it relates to science has been in the hands of social scientists, anthropologists, and other field-work based researchers. However, in recent years the dynamics of participation has changed dramatically to encompass a wide range of disciplines, including but not limited to architecture, design, planning, engineering, and medicine. Similarly, participation in policy has traditionally been solicited by large-scale organizations and agencies, but grassroots initiatives have also emerged as a driving force behind citizen initiated involvement in policymaking.

Examples of participatory processes on the ground are often highly specific, including a wide range of projects from participatory forest management, land cover evaluation, community natural resource planning, community-based facilities siting, large-scale displacement and resettlement projects, and poverty reduction efforts (Kunreuther et al. 1993; Mapedza et al. 2003; Mbile et al. 2003; Robiglio et al. 2003). Overall, participation is increasingly a global priority for countless types of projects. As a result, both the academic and applied work in this area are already vast and expanding.

What are the results of participatory efforts?

Although the need to evaluate and measure the success of both formal and informal participatory programs is widely acknowledged, coherent standards have yet to be widely implemented in practice. The specificity of both participatory research and embedded practice make it difficult to evaluate the extent to which different participatory tools and methods are effective relative to one another, especially when compared across a variety of contexts and applications. As a result, participation in practice currently suffers from the problem "when you have a hammer, all the world looks like a nail."

Because the call for participation is so overwhelming, participatory methods are often applied without clear theoretical underpinnings, goals, or even context within a larger project. Similarly, there are few standards for integrating participatory information into decision making, evaluating different levels of involvement, or defining success. Awareness of this growing gap has drawn attention to the importance of setting coherent standards for participatory methods, tools, and their applications, and underscoring the point that participation is not always necessary or appropriate. As Frank Fischer (2000) notes in his book <u>Citizens, Experts, and the Environment: The Politics of Local Knowledge</u>, "Citizen participation, in short, is a complicated and uncertain business that needs to be carefully thought out in advance."

Building Blocks of Participation

Given all the possibilities for public participation, how can citizens be effectively engaged and meaningfully involved in science and policy to support social justice? Based on the body of participation literature and applications outlined above, there is a growing need for better models to organize participatory projects and define general areas for measuring success in advance of implementation. One model, put forward here, is based on three main building blocks of participation: 1) information exchange (gathering and dissemination), 2) dialogue and stakeholder communication, and 3) decision making. These elements describe the basic levels of stakeholder interaction that form a large majority of participatory efforts. As Figure 6.1 illustrates, participation is a complex process consisting of a series of feedback loops among these three basic components. These building blocks do not individually or sequentially complete any participatory process; instead, effective participation requires an appropriate assembly of these fundamentals to address specific project goals and stakeholder needs.

Comparing this framework to Arnstein's (1969) ladder illustrates that, as participatory processes move up the levels of involvement on the ladder, combinations of building blocks are added to the process. Projects that are focused on persuasion and manipulation require only the most basic information dissemination, while even the lowest forms of consultation require some information gathering and feedback. The upper rungs of the ladder involve cyclic and iterative progressions

114

of various combinations of all three participatory building blocks. Although different levels of participation are associated with unique combinations of various building blocks, many participatory tools and methods focus solely on one of the three individual stages of participatory processes.

For example, household surveys and interviews, used frequently in social science studies, serve primarily as information gathering tools. Similarly, technical risk communications are often one-way information dissemination efforts. At a more detailed level of engagement, town meetings can serve as a medium for basic communication and dialogue among groups. Very few participatory tools focus on the dynamic, iterative decision making aspect of participation, and even fewer tools are designed to carry stakeholders through all the phases of participation and the resulting feedback loops.

The specialization of participatory tools and methods has aggravated the need for a simple, general structure to plan and evaluate participatory processes. As the English biologist Thomas Huxley stated, "The rung of a ladder was never meant to rest upon, but only to hold a man's foot long enough to enable him to put the other somewhat higher." Because a large majority of development planning and environmental decision making projects are inherently linked to spatial information, mapping has the potential to serve as a tool in support of all three building blocks across a variety of levels of participation. The next section outlines the dimensions of mapping and the strengths and weaknesses of spatial information to support participatory processes that move from one building block or one rung of Arnstein's ladder to another—linking science, policy, and social justice.

Location, Location, Location: Linking Mapping and Participation

Public participation in science and policy inherently has its basis in spatial information, where the locations of key resources and populations matter. This same link extends to social justice issues with key science and policy components. Craig and Elwood (1998) show how mapping has the potential to mobilize community groups, help organize communities around key issues, educate them about the details of the issue, plan potential responses, and communicate both social and technical concerns to policy makers, funders, and the wider public. As a result, various mapping methodologies have become particularly important for characterizing and understanding individual and community patterns of behavior and activities, their impacts on surrounding ecosystems, and related responses of local populations to environmental changes. Mapping has also become a major part of scientific studies and policy processes for visualization, analysis, and communication, particularly in the areas of development and environment. Two of the most widely used tools for these purposes are geographic information systems (GIS) technology and participatory mapping; however, these two tools have diverged in how they address the dynamic needs of a growing number of diverse projects.

GIS is a computer system and software capable of assembling, storing, manipulating, displaying, and analyzing geographically referenced data. The software, in tandem with related technologies like remote-sensing equipment and global positioning system (GPS) units, can be used to integrate layers of spatial data and perform complex analyses of the spatial relationships among objects and areas being mapped. While other maps represent a road simply as a line, a GIS has the potential to attach other information to the line and identify a significant cultural boundary or socioeconomic division between adjacent communities. In contrast, participatory mapping is a traditional method for collecting spatial information from community residents about their perceptions and relationships with local resources, places, or issues. The term participatory mapping, as it is used here, is defined very broadly as any combination of participation-based methods, such as interviews, for eliciting and recording spatial data (Goodchild 2007).

Specific examples of these methods include sketch mapping, scale mapping, and transect walking, among others. The maps resulting from these processes are highly specific to the cultures, languages, and education levels of different individuals and groups of participants acting as map makers. The final products range from maps drawn in the dirt with sticks and paper sketches to three-dimensional physical site models. The differences between these types of maps are akin to differences between a quick sketch drawn on the back of a napkin to give directions to a friend to a large-scale topographic map used by surveyors. The types and amounts of information displayed vary significantly. On the whole, both GIS and participatory mapping have important strengths for enabling participatory planning and decision making. However, the changing dynamics of participation coupled with some of the inherent limitations of each of these tools, has changed the roles that map makers and maps themselves traditionally play.

The change in who participates, when, and how often in a variety of science and policy efforts, has drawn attention to a wide range of information and communication technologies (ICTs) as potential tools to facilitate participatory development that is both inclusive and effective. However, the massive quantities and highly sophisticated presentations of data associated with many development and environmental projects have resulted in a divide beyond a lack of access to technology and even a lack of access to information. This new divide—between information and communication—is evident in a variety of scientific studies and policy outreach efforts, where various stakeholders and diverse groups require common information about a project, but understand and use this information very differently from one another. In some cases, information is both available and relevant, but it is represented in a form that is too general or too specific to be useful for the intended audience (Budhatoki et al. 2008).

GIS technologies provide one of the most striking examples of this paradox. The abilities of GIS to synthesize a wide variety of data and analyze complex spatial relationships have made it an essential planning tool for projects ranging from transport planning to forest conservation to infrastructure siting. As GIS technologies have been extended to more complex and diverse applications, the resulting maps and output from the system have also become increasingly intricate, and arguably, divergent from the users and communities the technology was originally intended to serve. This divergence has led to critical assessments of the social implications and applications of GIS and its outputs. In spite of these efforts and the rapid growth of new participatory GIS (PGIS) and public participation GIS (PPGIS) research areas, GIS technology and its maps remain largely focused on characterizing and analyzing attributes of locations, instead of the attributes of populations or livelihoods. With the increasing emphasis on environmental and social sustainability in development, GIS needs to move beyond conventional representations of *where* people live to describe more effectively the dynamics of *how* people live.

Since its inception, the potential of a GIS to simultaneously illustrate numerous aspects of a location has been its primary strength; however, with the emphasis on participatory information, this strength of the technology has also become a fundamental weakness of its output. GIS maps with multiple layers that include *all of the features* of a selected area, such as schools or green spaces, are now widely recognized as representing only one possible reality, and a collective reality at that. Rarely do all residents of a community interact with every school or park in their region, let alone in similar ways or for the same reasons. Individuals' connections with their physical surroundings are based on their priorities, perceptions, and preferences. In other words, populations are not homogenous, and *where* people live forms only a starting point for *how and why* they live there. This is central to the problem of ensuring social justice.

Although the overarching picture offered by GIS maps is important, this view is no longer enough. Effective development requires a disaggregation of both actual and perceived spatial relationships by gender, age, race, ethnicity, and income, among other characteristics, to understand and address the differential impacts of development among diverse populations. These impacts are widely acknowledged and studied in the growing critical GIS literature, but neither conventional nor participatory GIS currently serve the related information needs effectively. As Sieber (2004) and Cope and Elwood (2009) describe, the processes of data collection, integration, and map creation using GIS have only recently begun to change in response to the dynamics of development and environment related research and decision making.

In contrast, a variety of the existing methodologies for facilitating participation, such as participatory mapping, have emerged from different disciplines and been adapted to fill these gaps and promote more equitable decisions. These methods are promoted as counterparts to GIS for their ability to capture individuals' or groups' perceptions of local issues and development efforts. Although participatory maps, in contrast to GIS, describe *how* people live, many of these methods are limited in their usefulness. Often the process of data collection is extremely time-consuming, involving individual interviews and hand-drawn mapping, and the resulting information is difficult to compile and unwieldy for effective use by decision makers.

On the whole, the individual strengths and weaknesses of both participatory mapping and GIS are complementary. The next section describes the shared characteristics of these tools along

117

three dimensions. These dimensions form a theoretical basis for integrating spatial information across these methods to maximize their respective strengths and balance their weaknesses for facilitating participation in both science and policy with an eye to advancing social justice.

The Dimensions of Mapping

Traditionally, there has been little overlap between the users, audiences, and objectives of GIS and participatory mapping; however, with the recent changes in development practices, these two types of spatial information have gradually come together. Specialists in participatory methods and GIS have extended their respective research areas to include aspects of the other; but many of these efforts retain the strengths and weaknesses of their points of departure. For example, participatory GIS efforts typically retain the complexity and precision of a GIS, while hand-drawn maps input into GIS often remain largely informal and locally focused.

The growing movement toward integrating participatory methods and GIS highlights the fact that neither approach alone currently meets society's changing information needs. Combining participatory mapping methods and GIS and finding an appropriate balance between the two requires a clear assessment of their relative value for different applications. This assessment is essential for scientists, planners, and citizens alike. In spite of this awareness, there has been little critical analysis evaluating the effectiveness of current methods. This problem is not unique to mapping. As highlighted above, with the diversity of participation projects, their contexts, and their objectives, many participatory strategies have been applied in the absence of standard definitions and measures of success. Avoiding these indiscriminate applications of participatory tools, including mapping, requires a clear framework for planning and evaluation.

Both participatory mapping and GIS share three key dimensions that determine the balance between 1) social and spatial objectives, 2) accuracy and precision, and 3) representative and comprehensive displays of each type of spatial information. These attributes encompass how these two different mapping methods and their resulting maps are applied differently to engage citizens and communities in various science and policy contexts. Participatory mapping is typically socially-focused, locally accurate, and representative of participating map makers—the three elements of these dimensions focused on *how* people live. In contrast, the spatially-focused, geographically precise, and comprehensive characteristics of mapping are typically associated with GIS maps, describing *where* people live. The combination of GIS and participatory maps has the potential to balance these complementary attributes.

It is important to note here that the attributes along each dimension are not opposites nor are they exclusively associated with either GIS or participatory mapping. Instead these attributes illustrate the dominant values and objectives most commonly associated with each method. The interactions among them make up the unique characteristics of different maps and applications. Even within the domains of participatory mapping and GIS, there are varying emphases on these different characteristics. For example, certain types of participatory maps, based on transect

118

walks or scale mapping, demand far more spatial precision than others, such as sketch maps. Similarly, some GIS maps focus more strongly on social accuracy than others. For example, a map could represent a village as a single abstract point on a GIS layer or as a collection of polygons showing the dynamic changes in village boundaries, depending on the availability of relevant social data. Taken as a whole, the characteristics of maps along all three dimensions are dynamically driven by their underlying mapping methods and how the selected data is elicited, integrated, and displayed.

Social and spatial

In the case of the first dimension, the primary purpose of participatory maps is to elicit social information and organize it spatially; while GIS does the reverse, and arranges spatial information to shed light on social phenomena. This is not to say that GIS maps are not associated with social issues or vice versa, only that both GIS and participatory maps have different starting points, dominant characteristics, and influences. Collecting participatory information using traditional methods allows the focus of the dialogue to remain on social not spatial issues, while integrating the data into GIS formalizes the spatial characteristics and maximizes the relevance and potential for integration with other related data. Striking a balance between these two types of representations requires an integrated focus on both *how* and *where* people live, bringing both types of information together.

Accurate and precise

The terms accuracy and precision have specific scientific meaning; however, these two terms have almost become interchangeable in public parlance as they relate to mapping. In the context of mapping, the differences between the two are important, if subtle. The term accuracy, as it is used here, is intended to describe the correctness of information, while precision is a description of the resolution or "graininess" of the representation. In all cases it is important for maps to be both accurate and precise (to their respective scales and resolutions). In this case, most participatory mapping efforts focus on eliciting and recording accurate social information with varying degrees of spatial precision, while GIS maps demand a minimum degree of spatial precision to allow integration of social information with other widely used datasets, such as common administrative boundaries (e.g. states, counties, or neighborhoods). Ideally, all maps would be both socially and spatially accurate, as well as socially and spatially precise; however, this dimension is particularly important because early decisions in scientific studies on the required levels of precision or accuracy often drive how spatial data is collected and represented for related policy issues.

Representative and comprehensive

Participatory maps are largely subjective and focused on representing local perceptions and related descriptive information. As a result, these maps are often small-scale and widely understood, like a sketch map one would use to give directions based on familiar routes and landmarks. On the other hand, GIS maps are depictions based on comprehensive data, hence their visual complexity. The fundamentally different aims and applications of participatory mapping and GIS have shaped their dominant attributes. In theory, however, a collection of all possible locally representative views of a place could be assembled into a single comprehensive map, and participatory digital maps could both maintain the representative "lenses" or views provided by participatory maps while taking advantage of the comprehensiveness provided by collective integration in GIS.

Weighing the different attributes (deciding where a map should fall along each of these three dimensions) was once the work of cartographers and map makers who selected an appropriate projection and scale to display information in a clear manner for a given audience. As mapping technologies have become more and more accessible, individuals with new tools like Google Earth and other internet-based mapping software have moved into the role of map makers. Citizens and communities have been empowered with access to spatial information to display issues or problems as they see them, and not only to accept any maps or data that are presented to them. In this process, many of the decisions involved with defining an appropriate balance between social and spatial information, accuracy and precision, and representativeness and comprehensiveness have been made implicit, resulting in maps that are defined by available data and embedded assumptions in the mapping tool being used.

The attributes of a map that best describe a location to fit the needs of both the map makers and the map readers should drive the methods and objectives used to create the map. In other words, the mapping tools and approaches should not drive a map apart from the needs of the communities and issues to which it is being applied. For example, mapping is currently widely used for projects including border dispute resolution, resettlement planning, and community-based natural resource management. Each of these applications requires different levels of social and spatial information, accuracy and precision, and representative and comprehensive data. Defining the balance of attributes in advance of a project's implementation requires careful evaluation of the primary project and stakeholder needs. This framework is a critical tool for understanding how different mapping methods and their combinations of methods could be both best applied and best evaluated.

Overall, the differentiation between the two types of mapping in this chapter is not intended to suggest that both types are not used by communities and also useful to communities. This discussion and framework are simply intended to illustrate how taking advantage of the strengths of each type of map offers opportunities to better identify and address social justice issues, especially where science and technical concerns drive conversations toward spatially precise and comprehensive data systems that could ignore or push aside important socially relevant and locally representative information.

Implications for Science, Policy, and Social Justice

As maps and mapping tools have become more widely available to the public and more widely applied in science and policy, what are the implications for social justice? Given the complementary characteristics of participatory mapping and GIS described above, the synergies and conflicts between different types of spatial information are at the center of debates of how to better involve citizens in science and policy. Location determines whether someone is "in" or "out" of a study area, a project site, or a policy's reach. As a result, participation and inclusion are cornerstones of social justice, and mapping has emerged as a tool with multiple dimensions allowing for new audiences, users, and applications of scientific data. As the elaborate information storage and consolidation capacities of GIS are increasingly simplified to allow a wider variety of graphic displays and tailored to allow access by different users and audiences, mapping promises to become an increasingly important part of participatory processes. Below are three examples of new ways in which maps can be applied through science and social justice research to inform policymaking and advance participatory decision making, linking the building blocks of participation with the dimensions of mapping outlined above.

Reaching wider audiences with spatial information exchange

Environmental problems have grown dramatically in scale and scope in recent years. Problems such as acid rain have had impacts on extremely large areas and ecosystems, including the Adirondack and Appalachian Mountain regions. Although biophysical scientific assessments of the ecosystem impacts of acidification have improved, there is still significant uncertainty about the extent to which efforts to reduce the impacts of acidification will result in ecosystem improvements. Similarly, there is significant social scientific uncertainty surrounding such large-scale, ambiguously defined natural resources, where ecosystem boundaries do not correspond to traditional administrative boundaries, such as states. Developing policies to address environmental degradation depends on understanding how programs can be funded and implemented across these large regions. Given the spatial dimensions of the problem, there are two important areas where mapping can play an important role alongside economic policy studies: defining the extent of market (who cares about a natural resource) and the extent of resource (how big an area they care about) to elicit what people are willing to pay for environmental improvements.

In the case of the Appalachians, damage from acidification in the region is broad, but it is not clear whether residents of the region are particularly concerned about degraded resources in the states where they live, in neighboring states, on public lands, or more broadly across the region. In this case, mapping was used as part of a set of social science studies to gather information from individuals about the natural areas that they care about, the resources that they value, and their perceptions of environmental degradation. Figure 6.2 below illustrates the results of a mapping survey in this study that shows that national park or national forest boundaries can underestimate the extent of the resource that people value. Participants from both Virginia (Fig.

6.2a) and North Carolina (Fig. 6.2b) who were asked about natural areas and resources they value in the Appalachians overwhelmingly identified areas larger than corresponding national park boundaries for the Shenandoah National Park and the Great Smoky Mountain National Park.

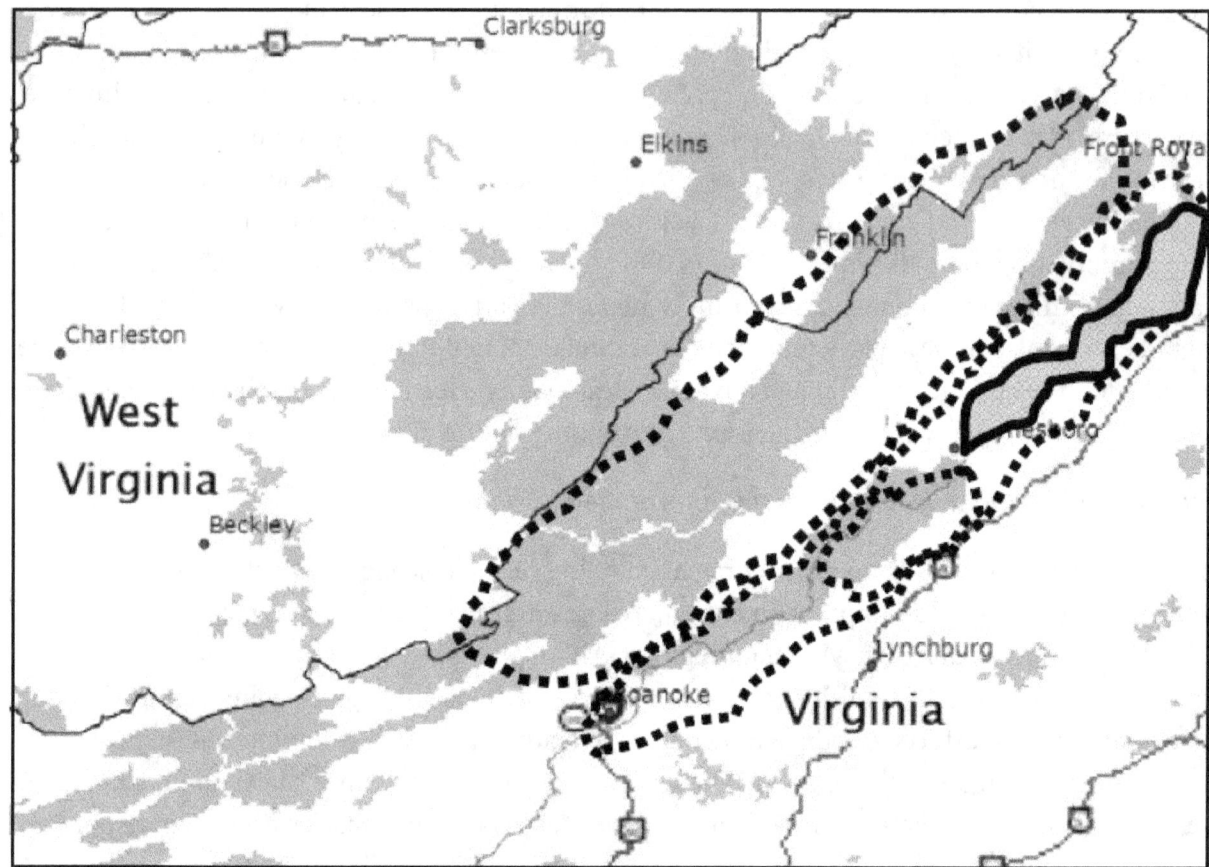

Fig. 6.2a. GIS map of Shenandoah National Park showing park boundary (solid black line) and "valued" resources from three mapping survey participants (dashed lines) (Vajjhala et al. 2008).

Bringing participatory maps into interdisciplinary scientific studies in geography, psychology and economics, among other fields, also has the potential to carry local priorities for new development and environmental investments forward into science-based policy, implicitly changing the nature of participation and inclusion in environmental decision making.

Using maps for communication and dialogue

Just as it is difficult to assess who values a natural resource and how much, it is also difficult to anticipate who considers themselves affected by proposed development projects or environmental changes. NIMBY (not-in-my-backyard) and an alphabet soup of related acronyms have become highly successful rallying cries for individuals and communities opposed to a variety of development planning and environmental management projects in recent years. In contrast, strategies to address public concerns have been far less effective. In spite of the growing popularity and use of mapping for both planning and participation purposes, little attention has been paid to the spatial elements of the NIMBY problem. In other words, how big is the backyard? Effectively siting facilities and accounting for significant heterogeneity of public opinion requires, first and foremost, a clear geographic characterization of the "backyard."

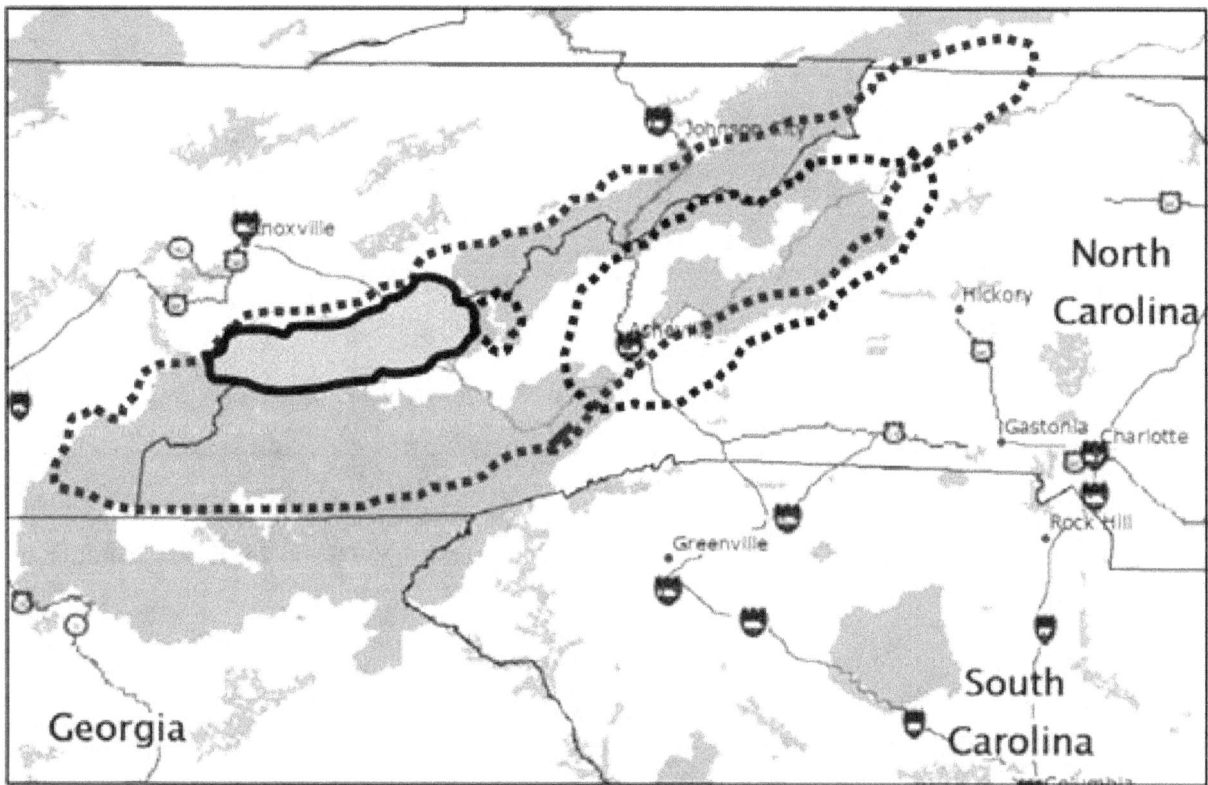

Fig. 6.2b. GIS map of Great Smoky Mountain National Park showing park boundary (solid black line) and "valued" resources from three mapping survey participants (dashed lines) (Vajjhala et al. 2008). (Color version is available in the EBook or from the author at StevenE@uvu.edu.)

Figures 6.3a and 6.3b from a mapping survey in Pittsburgh illustrate how even neighbors can have entirely different perceptions of and priorities for their shared community. The extent of the map maker's community in Figure 6.3a includes only the two blocks in all directions around his home. On the other hand, his neighbor's community in Figure 6.3b encompasses most of the places in the metropolitan region captured on her map. This is a central result that speaks directly to the concept of "backyard."

Mapping has the potential to elicit individual and community definitions of and values for their shared backyard using participatory methods. In parallel, local priorities, perceptions, and preferences can be integrated into GIS to evaluate planners' criteria for prospective sites alongside affected communities' priorities for new development. The iterative feedback and dialogue involved in siting decisions brings together the information gathering, dissemination, and exchange capacities of both participatory mapping and GIS, allowing for greater inclusion and flexibility in siting processes. The implication of this is enormous with the potential to shift siting and planning dialogues from an accept-or-oppose standoff to a meaningful negotiation.

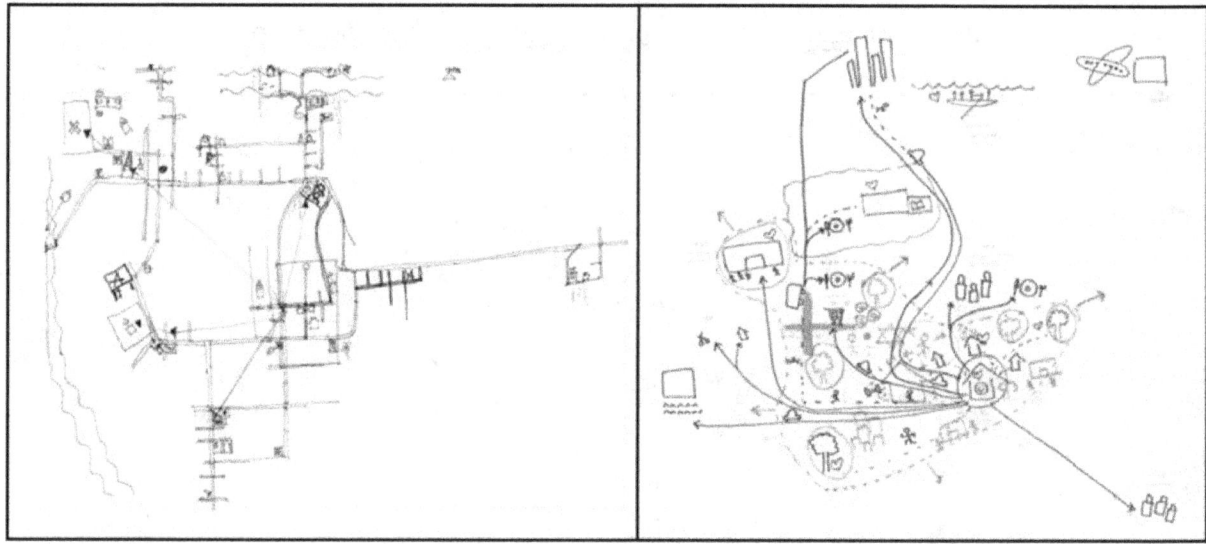

Fig. 6.3. Hand-drawn participatory sketch maps by two neighbors, a 74-year old man (left) and his 19-year old female neighbor (Vajjhala 2005). (Color version is available in the EBook or from the author at StevenE@uvu.edu.)

Cases of public opposition to major energy facilities reveal significant differences in individuals' priorities for their communities, preferences for different neighborhood attributes, and definitions of their backyards. Take for example, the now notorious Cape Wind proposal for a large-scale wind farm in Nantucket Sound, famously located in the Kennedy family's backyard. Not only does the geographic definition of "community" vary significantly among members of the same community, these perceived boundaries do not correspond with typical, artificial boundaries such as counties or zip codes. Given these variations, communicating with a broad audience requires a careful acknowledgement of stakeholders' diverse frames of reference in order to make development decisions that are socially just, locally relevant, understood, and accepted. Mapping has the potential to transform these dialogues, ranging from shifting national energy use patterns to supporting community-based natural resource management.

Generating maps to support participatory decision making

Although maps can help improve information exchange and foster dialogue as described above, maps, by themselves, do not make decisions. Instead, maps can play important decision support roles, providing better baseline data from which to evaluate issues at the intersection of the public, environmental and social sciences, and large-scale policies and regulations.

A final example at this intersection is risk communication. Major development plans and environmental decisions, like those involved in siting large dams and power plants, have grown progressively more complex. Project risks have become more long term (nuclear waste disposal), more uncertain (carbon sequestration), and more profound (development-induced displacement and resettlement programs). As the numbers of people affected and the potential impacts of a variety of projects have also multiplied, public awareness of different risks has evolved, and public opposition has become increasingly sophisticated. In response to this shift in public awareness, a variety of new methods have emerged to engage citizens and stakeholder groups in planning and decision making, particularly in the case of large-scale development-induced displacement and resettlement programs (Cernea and McDowell 2000). Similarly, the Gulf Coast hurricanes of 2005, which resulted in mass displacements from New Orleans and surrounding communities, link these extensively documented risks of resettlement to the larger issue of environmental justice (EJ) (Kurtz 2007). This is an area at the crossroads of science, advocacy, and policy, where mapping already plays a key role in identifying the existence and extent of environmental injustices. However, as EJ has increasingly become the focus of policy, mapping also has the potential to play a new decision support role. Figure 6.4 offers one example of how mapping can be used in EJ policy making, implementation, and evaluation. This map shows the locations of environmental justice small grants from the U.S. Environmental Protection Agency relative to the distribution of minority and low-income populations and environmental hazards based on the 2000 Census and Toxics Release Inventory, respectively.

More than a decade after the first EJ mandate was put in place with Executive Order 12898, the programs and initiatives associated with this order remain extremely broad in scope and intent. Compared to more traditional policies that establish environmental performance criteria or standards, EJ addresses both the process of making equitable decisions and their desired outcomes. Because federal EJ requirements are so general in nature (e.g., building community capacity, improving public awareness and environmental education, and expanding public participation), their implementation has been piecemeal and few systematic evaluations have been done of the implementation process or the major outcomes of federal EJ programs. Mapping has the potential to support EJ enforcement and program evaluation by integrating information on how the spatial distribution of EJ funds and projects has responded and changed in areas where minority and low-income communities live adjacent to environmental hazards.

Conclusion

The importance of science to social justice cannot be overstated. How social problems are identified, defined, and prioritized is intertwined with how these same problems are viewed through "objective" lenses, such as science. Although social justice is typically associated with advocacy, there is also a direct link between science and social transformation through policy. In all cases there is a need for better information and more informed decision making that takes into account differences in race, income, class, culture, and a host of other social characteristics that shape local priorities, perceptions, and preferences for new development and environmental policies. Mapping not only has the potential to make space for a "seat at the table" for underrepresented and underserved populations, but it also has the potential to expand the table altogether. In the words of Margaret Mead, "Never doubt that a small group of thoughtful, committed citizens can change the world. Indeed it is the only thing that ever has."

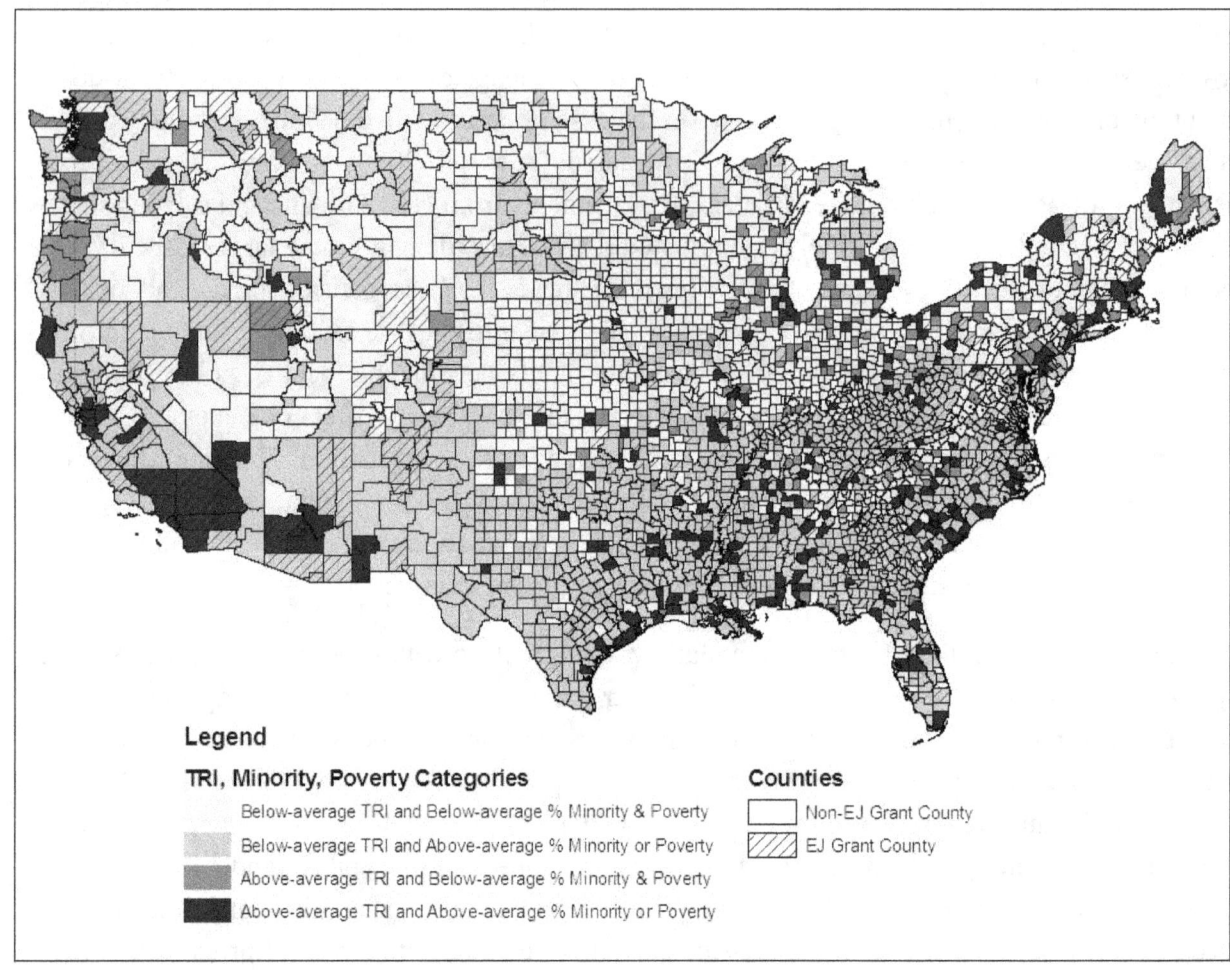

Fig. 6.4. Map of U.S. showing counties with both minority and low-income populations and toxic releases above the national average in dark brown and overlaying all counties that have received EPA EJ small grants with hatched lines (Vajjhala 2010). (Color version is available in the EBook or from the author at StevenE@uvu.edu.)

Disclaimer

This work is not a product of the United States Government or the United States Environmental Protection Agency, and the author is not doing this work in any governmental capacity. The views expressed are those of the author only and do not necessarily represent those of the US Government or the US EPA.

References

Abbot, J., R. Chambers, et al., 1998. Participatory GIS: Opportunity or oxymoron? PLA Notes, v. 33, pp. 27-33.

Arnstein, S. R., 1969. A ladder of citizen participation. Journal of the American Planning Association, v. 35, pp. 216-224.

Arvai, J. L., 2003. Using risk communication to disclose the outcome of a participatory decision-making process: Effects on the perceived acceptability of risk-policy decisions. Risk Analysis, v. 23, pp. 281-289.

Beierle, T. C. and J. Cayford, 2002. Democracy in Practice: Public Participation in Environmental Decisions. Resources for the Future, Washington, D.C.

Brodnig, G. and V. Mayer-Schönberger, 2000. Bridging the gap: The role of spatial information technologies in the integration of traditional environmental knowledge and Western science. The Electronic Journal on Information Systems in Developing Countries, v. 1, pp. 1-15.

Budhathoki, N. R., B. Bruce, and Z. Nedovic-Budic, 2008. Reconceptualizing the role of the user of spatial data infrastructure. GeoJournal, v. 72, pp. 149-160.

Burke, E. M., 1968. Citizen participation strategies. Journal of the American Institute of Planners, v. 34, pp. 287-294.

Cernea, M. M. and C. McDowell (eds.), 2000. Risks and Reconstruction: Experiences of Resettlers and Refugees. The World Bank, Washington D.C.

Chambers, R., 1983. Rural Development: Putting the Last First. Pearson Education Limited, Edinburgh Gate, Essex, UK.

Chambers, R., 1994. The origins and practice of participatory rural appraisal. World Development, v. 22, pp. 953-969.

Chess, C. and K. Purcell, 1999. Public participation and the environment: Do we know what works? Environmental Science and Technology, v. 33, pp. 2685-2692.

Cope, M. and S. Elwood (eds.), 2009. Qualitative GIS: A Mixed-methods Approach. Sage, London.

Cornwall, A. and R. Jewkes, 1995. What is participatory research? Social Science Methods, v. 41, pp. 1667-1676.

Craig, W. J. and S. A. Elwood, 1998. How and why community groups use maps and geographic information. Cartography and Geographic Information Systems, v. 25, pp. 95-104.

Craig, W. J., T. M. Harris, and D. Weiner (eds.), 2002. Community Participation and Geographic Information Systems. Taylor and Francis, London.

Davis, J. and D. Whittington, 1998. "Participatory" research for development projects: A comparison of the community meeting and household survey techniques. Economic Development and Cultural Change, v. 47, pp. 73-94.

Dunn, C. E., P. J. Atkins, and J. G. Townsend, 1997. GIS for development: A contradiction in terms? Area, v. 29, pp. 151-159.

EPA (U.S. Environmental Protection Agency), 2005. Public Involvement Case Studies, EPA Online Publications Archive. Available online at http://www.epa.gov/publicinvolvement/casestudies.htm.

Fiorino, D. J., 1990. Citizen participation and environmental risk: A survey of institutional mechanisms. Science, Technology, and Human Values, v. 15, pp. 226-243.

Fischer, F., 2000. Citizens, Experts, and the Environment: The Politics of Local Knowledge. Duke University Press, Durham, NC.

Goodchild, M. F., 2007. Citizens as sensors: the world of volunteered geography. GeoJournal, v. 69, pp. 211-221.

Inhaber, H., 1998. Slaying the NIMBY Dragon. Transaction Publishers, New Brunswick, NJ.

Irvin, R. A. and J. Stansbury, 2004. Citizen participation in decision making: Is it worth the effort? Public Administration Review, v. 64, pp. 55-65.

Kasperson, R. E., G. Berk, D. Pijawka, A. B. Sharaf, and J. Wood, 1980. Public opposition to nuclear energy - retrospect and prospect. Science, Technology & Human Values, v. 31, pp. 11-23.

Kienberger, S., F. Steinbruch, A. Luis, A. Gomes, and T. Blaschke, 2005. The potential of community mapping and community integrated GIS: A study in the Sofala Province, Mozambique. In 10th International Conference on Information & Communication Technologies (ICT) in Urban Planning and Spatial Development & Impacts of ICT on Physical Space, M. Schrenk (ed.), Vienna, Austria, pp. 697-703.

Kierkegaard, S., H. V. Hong (trans.), and E. H. Hong (trans.), 1998. The Point of View. Princeton University Press, Princeton, NJ.

Kunreuther, H., K. Fitzgerald, and T. D. Aarts, 1993. Siting noxious facilities: A test of the Facilities Siting Credo. Risk Analysis, v. 13, pp. 301-318.

Kurtz, H. E., 2003. Scale frames and counter-scale frames: constructing the problem of environmental injustice. Political Geography, v. 22, pp. 887-916.

Kurtz, H. E., 2007. Environmental justice, citizen participation and Hurricane Katrina. Southeastern Geographer, v. 47, pp. 111-113.

Mapedza, F., J. Wright, and R. Fawcett, 2003. An investigation of land cover change in Mafungautsi Forest, Zimbabwe, using GIS and participatory mapping. Applied Geography, v. 12, pp. 1-21.

Mbile, P., A. DeGrande, and N. Okon, 2003. Integrating participatory resource mapping and geographic information systems in forest conservation and natural resources management in Cameroon: A methodological guide. The Electronic Journal on Information Systems in Developing Countries, v. 14, pp. 1-11.

McCall, M. K., 2003. Seeking good governance in participatory-GIS: A review of processes and governance dimensions in applying GIS to participatory spatial planning. Habitat International, v. 27, pp. 549-573.

McDaniels, T. L., R. S. Gregory, and D. Fields, 1999. Democratizing risk management: Successful public involvement in local water management decisions. Risk Analysis, v. 19, pp. 497-510.

NRC (National Research Council), 1996. Understanding Risk: Informing Decisions in a Democratic Society. National Academy Press, Washington, D.C.

Pretty, J. N., 1995. Participatory learning for sustainable agriculture. World Development, v. 23, pp. 1247-1263.

Renn, O., T. Webler, and P. M. Wiedemann (eds.), 1995. Fairness and Competence in Citizen Participation: Evaluating Models for Environmental Discourse. In Technology Risk and Society: An International Series in Risk Analysis. Kluwer Academic Publishers, Dordrecht, The Netherlands.

Robiglio, V., W. A. Mala, and M. C. Diaw, 2003. Mapping landscapes: Integrating GIS and social science methods to model human-nature relationships in southern Cameroon. Small-scale Forest Economics, Management and Policy, v. 2, pp. 171-184.

Rowe, G. and L. J. Frewer, 2000. Public participation methods: A framework for evaluation. Science, Technology & Human Values, v. 25, pp. 3-29.

Sieber, R. E., 2004. Rewiring for a GIS/2. Cartographica: The International Journal for Geographic Information and Geovisualization, v. 39, pp. 25-39.

Tripathi, N. and S. Bhattarya, 2004. Integrating indigenous knowledge and GIS for participatory natural resource management: State-of-the-practice. The Electronic Journal on Information Systems in Developing Countries, v. 17, pp. 1-13.

Vajjhala, S. P., A. Mische John, and D. A. Evans, 2008. Determining the extent of market and extent of resource for stated preference survey design using mapping methods. RFF Discussion Paper 08-14. Resources for the Future, Washington, D.C.

Vajjhala, S. P., 2005. Mapping Alternatives: Facilitating Citizen Participation in Development Planning and Environmental Decision Making. Ph.D. Thesis, Carnegie Mellon University.

Vajjhala, S. P., 2006. "Ground truthing" policy: Using participatory map-making to connect citizens and decision makers. Resources, Summer (162). Available online at http://www.rff.org/focus_areas/features/Pages/Participatory-map-making.aspx.

Vajjhala, S. P., 2010. Building community capacity? Mapping the scope and impacts of the EPA Environmental Justice Small Grants Program. Research in Social Problems and Public Policy, v. 18, pp. 353-381.

Webler, T., 1999. The craft and theory of public participation: A dialectical process. Journal of Risk Research, v. 2, pp. 55-71.

World Bank, 1995. World Bank Participation Source Book. Environmental Department Papers, World Bank, Washington, D.C.

Yapa, L. S., 1991. Is GIS appropriate technology? International Journal of Geographic Information Systems, v. 5, pp. 41-58.

Chapter 7: Using Scientifically Oriented School-Based Projects to Further Environmental Justice

Nicky Sheats, Theodore Carrington, Fletcher Harper, Kim Gaddy, and Valorie Caffee

Summary

In this chapter we argue that scientifically oriented school-based projects have the potential to empower student and adult residents of low-income and/or Of Color communities to achieve environmental justice goals. The chapter initially reports on a project called the New Jersey Urban Air Quality Education and Awareness Initiative (also referred to as the Initiative) that was implemented by the New Jersey Environmental Justice Alliance (also referred to as the Alliance) and several partners in 2005. It then describes an expanded form of this project the Alliance hopes to implement in the near future and presents general characteristics that should be part of any scientifically oriented school-based project that is intended to empower a community. We use environmental justice to provide the social justice framework for the chapter.

The first Initiative involved high school students from Camden, Trenton, and Newark, New Jersey, in monitoring airborne particulate matter (PM) concentrations in their respective cities. A second expanded Initiative will perform more purposeful monitoring, be integrated more fully into the school curriculum, involve students in policy development, and ask students to conduct a community education campaign that will empower neighborhood residents to address the problem of PM air pollution.

Community empowerment and community-based participatory research are also discussed in this chapter in order to provide a context for the general characteristics that we believe all scientifically oriented school-based projects should have. We end the chapter by suggesting that if our nation educated, graduated and employed more scientists Of Color, there would be more scientific projects that had environmental and social justice implications.

Introduction

This chapter suggests a manner in which science can be used to further social justice goals by discussing a specific project that had a positive social influence when it was implemented the first time, and has the potential to significantly benefit low-income neighborhoods and Of Color communities when it is repeated in an expanded format. The project, the New Jersey Urban Air Quality Education and Awareness Initiative, involved high school students from Camden, Trenton and Newark, New Jersey, in airborne particulate matter (PM) monitoring in their respective communities. Details of this initial project and plans for another expanded form of this endeavor are discussed in this chapter. The expanded form of the project is also placed in the context of community-based participatory research, a research approach that involves community residents in each stage of the development and implementation of a research project.

The social justice framework for the chapter is provided by the field of environmental justice, which emerged as an area of concern in both the social justice and environmental worlds in the late 1980s. An early classic description of the environmental justice field is that it addresses the concerns held by environmental justice advocates, and substantiated by a number of investigations, that low-income neighborhoods and communities Of Color often suffer from a disproportionate burden of pollution due to environmental hazards siting decisions that are based at least partly on race and class (GAO 1983; Mohai and Bryant 1987, 1992; Pastor et al. 2007). The U. S. Government Accounting Office (GAO 1983) and Mohai and Bryant (1987) produced seminal reports that helped demonstrate the correlation between race, class and the siting of environmental hazards. Along with a book written by sociologist Robert Bullard (1990), entitled Dumping in Dixie, these early reports also helped establish environmental justice as a public issue that needed to be addressed. In 2007 the United Church of Christ report was repeated (Bullard et al. 2007) and the troubling relationship between race, class and the location of environmental hazards was reaffirmed. More expansive definitions of the discipline became appropriate early in its development and recognized a more holistic view of environmental justice by including issues related to transportation, planning, development, employment, education and inclusive decision-making (Bullard et al. 2007).

In this chapter, we argue that scientifically oriented school-based projects, such as an expanded PM monitoring program, have the potential to further environmental justice by empowering students and other residents of neighborhoods that are suffering from a disproportionate amount of pollution. To make this assertion convincing, we also present information that provides background and a context for these endeavors. Immediately following this introduction, community empowerment is defined and different aspects of this concept are explored. The next section contains a similar treatment for community-based participatory research. This research model is defined and discussed. This section also has brief discussions of community-owned and managed research, in addition to service-learning, which are, respectively, a research model and a programmatic concept that involve students in community activities that have relevance for scientifically oriented school-based projects. The next two sections of the chapter discuss the initial monitoring project, which was implemented by the New Jersey Environmental Justice Alliance and several partners, and the elements of an expanded monitoring project. A short discussion of previous community-based monitoring projects conducted by other organizations follows these sections and then general principles for scientifically oriented school-based projects are presented. The chapter concludes with discussion of a systemic problem that we believe negatively impacts the number of scientific projects that have social justice implications: the low number of scientists Of Color that are educated each year in our country.

Community Empowerment

One of the most important aspects of an expanded monitoring project would be its potential to empower the community in several ways. However, before we discuss how this could be achieved we will define and discuss empowerment in general. Empowerment has been defined in different ways (Israel et al. 1994; Page and Czuba 1999), especially across disciplines (Israel et al. 1994) and in different situations, or with different populations (Perkins and Zimmerman 1995; Page and Czuba 1999). For purposes of our expanded monitoring project, we favor the following definition offered by Perkins and Zimmerman (1995): "an intentional ongoing process centered in the local community, involving mutual respect, critical reflection, caring and group participation, through which people lacking an equal share of valued resources gain greater access to and control over those resources" (quoting a 1989 definition by the Cornell Empowerment Group). This definition is appealing because it explicitly mentions group participation as part of the empowerment process and locates the process, both literally and figuratively, in the local community. But we recognize that others may favor the simplicity of a shorter, more straightforward, definition offered by the same authors: "… a process by which people gain control over their lives, democratic participation in the life of their community (citation omitted), and critical understanding of their environment (citations omitted)."

Both definitions have merit and several observations should be made about them. First, both define empowerment as a process (also see Page and Czuba (1999) and Laverack and Wallerstein (2001)), however it can also be viewed as an outcome (Perkins and Zimmerman 1995; Laverack and Wallerstein 2001). Second, although thus far we have discussed empowerment in terms of the community, it can also occur on the individual level (Israel et al. 1994) and, in fact, some would say that it must occur in individuals before it can be achieved on the community level (Page and Czuba 1999; Cook 2008).

In the sections of the chapter that discuss the initial monitoring project and plans for an expanded project, specific examples will be given that illustrate the terms and ideas introduced above.

The Community-Based Participatory Research Framework

The general characteristics of a scientifically oriented school-based project, and the proposed expanded monitoring project itself discussed later in this chapter, present a format for creating a type of research project that fully involves local high school students and other community residents. Research that extensively involves community members is called participatory research and has been used increasingly in public health and related fields due to its potential to improve health and reduce health disparities (Israel et al. 2005a; Cargo and Mercer 2008). Because it generally attempts to involve community residents in most, if not all, aspects of research projects, we believe that it can also result in community empowerment. Various types of participatory research approaches exist (Israel et al. 1998; 2005a; Cargo and Mercer 2008), but one specific type of participatory research that has been used by the environmental justice community, and been viewed positively by environmental justice advocates (Shepard et al. 2002), is community-based participatory research (CBPR). Cook (2007) defines CBPR as: "…essentially an egalitarian and action-oriented endeavor that strives to equitably involve community members, organizational representatives and researchers in all aspects of the research process." (See also Israel et al. (2005a)). Aspects of the research collaboration include forming research questions, designing the study, collecting data, interpreting study results and communicating research results to the broader community (Shepard et al. 2002; Israel et al. 2005a). This broader community could consist of community residents, media and policymakers (Shepard et al. 2002). In order to conduct effective CBPR, O'Fallon and Dreary (2002) recommend adhering to the following six principles, which are also endorsed by the National Institute of Environmental Health Sciences: 1) promote active participation and collaboration on an equal basis by all partners at all stages of the research; 2) foster co-learning by enabling both researchers and community residents to contribute their respective expertise to the research, and by creating an atmosphere where partners can learn from each other; 3) ensure research is community driven by having the project "guided by the environmental health issues or concerns of the community"; 4) communicate results to all partners in a manner that is understandable, culturally appropriate, and respectful to everyone; 5) ensure that the research and any intervention strategies are culturally appropriate through active participation of community residents at the earliest stages of the project; and 6) allow "community to be defined by the people whose health is most likely to be affected by the research."

These general principles are not set in stone and other authors have presented CBPR principles that are similar but not exactly the same as those detailed above (Israel et al. 1998, 2005a, 2005b). A close variation to CBPR that could provide an alternative research approach in some situations is the community owned and managed research (COMR) model. This approach provides more control to the community because it mandates that a community-based organization must be the project's principal investigator and therefore manage the research and receive the funding (Heaney et al. 2007; Wilson et al. 2007). In contrast, in the CBPR research model, university researchers are usually the principal, or at least co-principal, investigators (Heaney et al. 2007) and therefore have more control over the research, relative to the community, than in COMR. However, it is important to note that the COMR model is not appropriate in all situations that lend themselves to CBPR because it requires the community-based organization that manages the research to have a high level of organizational capacity (Heaney et al. 2007; Wilson et al. 2007). In situations where there is a community-based organization with the required level of organizational capacity available to participate in a potential research project, utilization of the COMR model should be given strong consideration since it ensures not only community participation in the project but also community decision-making authority. In our own work, instead of using COMR as a framework, we have chosen to develop our own close variation of CBPR specifically for scientifically oriented school-based research projects. We do so because we believe that tailoring a research model to this special category of research projects will provide them with the best chance of success.

Service-learning could provide another close analogy for the research model we propose later in this chapter. In the service-learning model, students, usually in high school or college, actively participate in an activity that benefits the community (Billig 2000; Butin 2003). However, we believe that CBPR provides a better context for the work we are advocating here because it emphasizes community involvement in research whereas service-learning focuses on helping the community in a manner that can include activities other than research. We envision scientifically oriented school-based projects as also being research oriented. Both the type of activity, research as opposed to other activities, and the nature of the interaction with the community, community involvement as opposed to community service, lead us to favor CBPR over service-learning as a model for our work (Ward and Wendel 2000; O'Fallon and Dreary 2002; Shepard et al. 2002; Butin 2003; Israel et al. 2005a, 2005b). But service-learning could provide a viable model for other types of community oriented research. General characteristics for scientifically oriented school-based projects are presented later in this chapter.

New Jersey Urban Air Quality Education and Awareness Initiative

The New Jersey Environmental Justice Alliance (also referred to as the Alliance) was the primary organization that planned and implemented the initial monitoring project, which was called the New Jersey Urban Air Quality Education and Awareness Initiative (also referred to as the Initiative). The Alliance was formed in early 2003 and includes several dozen organizations in its membership. It is the only statewide environmental organization in New Jersey that focuses solely on environmental justice issues and is further distinguished from other environmental organizations because a majority of both its membership and leadership are people Of Color. In 2004, the Alliance was approached by the New Jersey Department of Environmental Protection (NJDEP) to provide advice on legislation it was helping to develop that would reduce emissions of diesel PM. The legislation was eventually fully developed by NJDEP and adopted by the state legislature as the Diesel Retrofit Law (P.L. 2005, c. 219 and P.L. 2006, c.94, N.J.S. 26:2C-8.26 et seq.).

Diesel PM is just one type of fine PM air pollution (Schneider and Hill 2005). Fine PM refers to all airborne PM that is 2.5 microns or less in diameter (Godish 1997), and almost certainly causes tens of thousands of premature deaths in the United States annually as evidenced by the facts that it has been estimated to cause 14,000 to 24,000 deaths in California (California Environmental Protection Agency 2009), and at least 1900 in New Jersey each year (NJDEP 2008). It is also associated with an alarming number of illnesses and physical disorders such as cardiovascular disease (Pope et al. 2004; Jerrett et al. 2005) and pulmonary disorders that include lung cancer (Dockery et al. 1993; Pope et al. 2002; Jerrett et al. 2005), asthma (NJDEP 2008) and diminished lung function in children (Drunekreef et al. 1997). Fine PM is an environmental justice issue because people Of Color and low-income community residents are almost certainly dying, and becoming ill, from this pollutant at a higher rate than other segments of our country's population since they live in disproportionate numbers in urban areas (Massey and Denton 1993) where fine PM concentrations are typically the highest (EPA 2005).

The Alliance engaged NJDEP and New Jersey legislators on the details of the diesel emissions reduction legislation and went a step further by making recommendations on other issues related to reducing fine PM concentrations in urban areas in the state (Sheats et al. 2008). One recommendation the Alliance made repeatedly was that a community-based PM monitoring system should be established within New Jersey. This policy recommendation was prompted by concern within the Alliance and communities in several New Jersey cities that there were an insufficient number of monitors operating in the state to adequately capture the variation in fine PM concentrations that might occur within urban areas. At the time, NJDEP was maintaining fine PM monitors in 22 locations in the state (NJDEP 2005a); a number that averages to approximately one monitor per county since New Jersey has 21 counties. Alliance members felt the optimal solution was to place additional monitors in the hands of community members in the hope that knowledge of fine PM concentrations in their neighborhoods would increase their desire to understand the detrimental health effects associated with this pollutant, and eventually lead to actions intended to reduce its concentration in their community. An opportunity to work on this issue at the community level presented itself when the John S. Watson Institute for Public Policy of Thomas Edison State College (also referred to as the Watson Institute), which works on public policy issues faced by New Jersey residents and is located in central New Jersey, approached the Alliance with the possibility of forming a community-college partnership to implement an educational project that focused on diesel and fine PM. The Alliance suggested that the project take the form of airborne PM monitoring by high school students in various parts of the state. There were several reasons the Alliance chose to work with high school students. One was because its members believed that, as Loh et al. (2002) stated about a monitoring project in a Boston neighborhood that had extensive student involvement, "Youth are often the most impassioned and articulate spokespeople for community issues." By working with youth, we also saw an opportunity to create and partially train a new generation of environmental justice leaders and advocates. By working with schools we thought there might be the possibility of institutionalizing and spreading the monitoring project across the state since schools exist in every community. If the monitoring project were replicated over a consistent time period in a sufficient number of schools, this could lead to the creation of a state-wide community-based and community-driven monitoring system. The Watson Institute accepted the idea of working with students and schools, and planning on the "New Jersey Urban Air Quality Education and Awareness Initiative" began. In addition to providing a sound educational experience for students, the Alliance also hoped the monitoring project might empower the students and other members of the community to act on air pollution issues, and prove that lay people could effectively operate the monitors employed by the project.

Even before approaching the Alliance, the Watson Institute had a promise of funding from the Ford Foundation, and so the two largest obstacles to project implementation that remained were obtaining the cooperation of several high schools and determining how to accomplish the monitoring using the existing budget. A full-time administrator was hired to coordinate the work with the schools and the Clean Air Task Force (CATF), a national non-governmental organization that performs advocacy work intended to reduce diesel emissions, was contacted to help perform the monitoring. CATF had experience in conducting air pollution monitoring with lay people utilizing the DustTrak and Ptrak, handheld real-time monitors. The DustTrak measures concentrations of fine PM by determining the mass of the particles (Chung et al. 2001; Levy et al. 2001; Yanosky et al. 2002) while the Ptrak determines the concentrations of ultrafine PM, airborne particulates less than 0.1 microns in diameter (Kittelson 1998), by counting the number of particles (TSI 2011). The CATF provided the monitors, trained all project participants in their use, and provided onsite assistance at the time of monitoring.

The New Jersey Urban Air Quality Education and Awareness Initiative occurred in the spring of 2005 and involved high school students in monitoring airborne PM concentrations in Camden, Trenton and Newark, New Jersey. The students from Newark and Camden were members of science classes, and three of four Trenton students were part of an environmentally oriented mentoring program. The monitoring occurred over several days and the students were assisted by adult members of both the Alliance and the Clean Air Task Force. Prior to monitoring, educational sessions were conducted with the students on: 1) environmental justice and related issues, and 2) the scientific aspects of PM air pollution and issues related to it. Training sessions on the use of the monitoring equipment were also conducted for the students and project-planning sessions were held where students helped select monitoring sites and times. Although the number of students that actively participated in the monitoring was a relatively modest sixteen, additional students and members of the public received information on the Initiative when the students from Trenton made presentations at a countywide science fair and to elementary school classes at different Trenton schools. The public was also reached by the Initiative through television, newspapers and radio media that disseminated information on the project after press conferences held in each city at which the students spoke about their monitoring activities and the dangers of PM air pollution.

Shortly after the conclusion of the Initiative, a scientifically trained member of the Alliance performed simple statistics on the data and presented them in the form of figures and tables (Sheats 2005). The data revealed several monitoring sites where concentrations in excess of the 15.0 $\mu g/m^3$ federal annual standard for fine PM (40 CFR section 50.7(a)) were recorded (Table 7.1).

Teaching student participants about environmental justice, and educating the students and public about PM air pollution, were two of the most important accomplishments of the Initiative. Another major accomplishment was demonstrating that high school students and lay adults, under supervision from scientifically trained adults, could utilize the DustTrak and Ptrak instruments to perform community-based air monitoring to produce, what we believe, is scientifically credible data. Before the project began there was a real question, especially on the part of NJDEP personnel, whether the Initiative could yield scientifically credible data or whether it would be "just" an educational program. In this context, by credible the Alliance meant data that were sufficiently trustworthy to warrant NJDEP using its own universally acknowledged reliable monitoring equipment to measure PM concentrations in an area where our monitoring found high PM levels. It also meant that data produced by the project could at least be used to say that one monitoring location had significantly higher or lower PM concentrations than another location, even if the values of the recorded concentrations were not deemed to be exact.

Table 7.1. Original and adjusted DustTrak PM2.5 concentrations ($\mu g/m^3$) recorded by the New Jersey Urban Air Quality Education and Awareness Initiative. Least squares regression equation used to adjust concentrations was $y = 6.3188 + 0.23159x$.

Location	Date	Start Time	Sample Duration	Mean PM2.5 ($\mu g/m^3$)	Adjusted PM2.5 ($\mu g/m^3$)
Trenton					
North side, Prospect & Olden	4/7/05	13:58:11	0:20:30	42	15
South side, Prospect & Olden	4/7/05	13:55:36	0:25:10	43	16
Upwind, Princeton & Olden	4/7/05	14:30:34	0:27:40	47	16
Downwind, Princeton & Olden	4/7/05	14:30:32	0:27:40	45	16
Camden					
Morgan Blvd & Broadway	5/18/05	9:35:12	0:30:50	27	12
Westside of Broadway	5/18/05	10:26:39	0:27:00	23	11
Liney Ditch Park Comm., DustTrak 2	5/18/05	11:15:42	0:44:00	24	11
Liney Ditch Park Comm., DustTrak 1	5/18/05	11:13:58	0:46:30	26	12
Waterfront South Comm.	5/18/05	9:35:26	0:31:10	18	10
Walter Rand Bus Term., Outdoors	5/17/05	9:35:01	0:32:30	32	13
4th St. & Jefferson	5/17/05	10:30:27	0:29:40	24	11
East Atlantic & 9th St.	5/17/05	13:17:24	0:15:50	29	13
West Atlantic & 4th St.	5/17/05	13:44:24	0:17:40	32	13
Walter Rand Bus Term., Indoors	5/17/05	9:36:30	0:33:00	65	20
Broadway & Winslow	5/17/05	10:30:27	0:29:30	25	12
Broadway & Atlantic	5/17/05	13:17:41	0:31:50	35	14
Newark					
Frelinghuysen Comm., DustTrak 1	5/12/05	9:46:59	0:30:20	7	8
Press Event, DustTrak 1	5/12/05	10:44:10	0:43:30	12	9
Frelinghuysen Comm., DustTrak 2	5/12/05	9:52:36	0:32:00	10	8
Press event, DustTrak 2	5/12/05	10:49:50	0:43:30	12	9
Raymond Blvd. & Lockwood	5/11/05	10:22:37	0:57:40	75	22
Forest Hill	5/11/05	12:51:07	0:56:40	62	20
North 6th St. & Park Ave.	5/11/05	14:12:39	0:10:10	71	21
Forest Hill, Grafton & Branchwood	5/11/05	12:56:30	0:41:50	71	21
Raymond Blvd. & Blanchard	5/11/05	10:23:51	0:51:40	88	25

There were two primary issues with respect to data credibility. The first was whether high school students and adult lay project participants could use the machines properly and perform other aspects of monitoring in a manner that would produce reliable data. Although it can never be ensured that all aspects of any scientific endeavor are performed correctly, we believe the training sessions on the use of the machines, the user-friendly nature of the machines, the planning sessions on other aspects of the project, the fact that a scientifically trained project participant was present at each monitoring location and an adult was always present with a student, all combined to foster the production of credible data. The presence of adults with the students at all times during the project not only helped make sure that the machines were used properly and reduced human error but we feel should allay any fears that students falsified data or behaved unethically in any way during the project. It was also important that one of the adults present in Camden and Newark was the students' teacher with whom they obviously had a relationship that predated the monitoring project and therefore provided the students with a sense of trust, confidence and responsibility.

The second issue regarding the credibility of the data involved the reliability of the DustTrak itself because previous investigations that utilized the machine had determined that the instrument generally overestimated ambient fine PM concentrations by a factor of two to three (Chung et al. 2001; Levy et al 2001; Yanosky et al. 2002). In order to gather data that could be used to correct this bias, one of the DustTrak monitors used during the Initiative was co-located for almost four days with a more precise PM monitor called a Tapered Element Oscillating Microbalance (TEOM) that was maintained by NJDEP. A least squares regression performed on data produced by the TEOM and DustTrak showed that the DustTrak did tend to overestimate concentrations but it also demonstrated a relatively high degree of correlation between the two data sets with $R^2 = 0.81$ (Sheats 2007). This indicates that fine PM concentrations measured by the DustTrak are reliable if they are corrected with a regression equation produced by a calibration between the DustTrak and another type of monitor that has demonstrated a high degree of accuracy, especially since other investigators have also successfully performed this type of calibration involving a DustTrak in more detail and using more data (Chung et al. 2001; Yanosky et al. 2002). However, it should be noted that there still remain some points of contention on this topic between the Alliance and NJDEP, because NJDEP staff still harbored reservations concerning the performance of the DustTrak despite the calibration that was conducted (NJDEP, personal communication).

Demonstrating that data produced by the DustTrak was reliable was a higher priority than accomplishing the same task for the Ptrak because there are federal health-based standards for fine PM (40 CFR section 50.7(a)), whereas no such standard exists for ultrafine particles. The standard provides a benchmark that can be used to determine whether local fine PM concentrations are at unhealthy levels.

At the conclusion of the Initiative, Alliance participants felt that one of the original goals of the project, empowering the community through education so they could act on the issue of PM air pollution, had been at least partially achieved. The student participants received a sound learning experience on PM air pollution, and the broader field of environmental justice, through the classroom work conducted prior to the monitoring and by actually visiting urban neighborhoods and recording PM concentrations. The social consciousness of the students, especially those from Camden, had also been enhanced by their Initiative experiences. For example, in the southern New Jersey city of Camden, much of the monitoring had been performed in a community called Waterfront South. There is a large county sewage plant, a cement plant, a licorice factory, an incinerator, and numerous smaller polluting businesses in, or near, the neighborhood (NJDEP 2005b). As a result of the high concentration of stationary polluters, a noticeable smell often permeates the neighborhood. Although the students who monitored in this neighborhood were from Camden, they were surprised by what they saw and smelled. All of the students planned to attend college and at the conclusion of the Initiative, several commented that they had not intended to return to Camden after college, but after seeing the needs of the area's communities, they were now considering remaining in the city and assisting its residents. We believe that the knowledge gained by the students and the consciousness change they experienced regarding their relationship to their community is an example of the individual empowerment discussed above. The monitoring by the students and the press conferences they held at the end of monitoring sessions exemplify community empowerment. These student-initiated actions were performed to obtain additional knowledge of a critical issue affecting their community, to raise community and public awareness of the issue, and to educate community stakeholders and the public about the issue. This is an example of community empowerment as a process. Ultimately, the organizers of the Initiative felt that the scientifically oriented educational project not only demonstrated that lay people could use scientific equipment, but also used science as a vehicle to introduce students to a critical environmental health issue in their communities, and engender a sense of social obligation to their home cities.

However, the organizers also thought that more could be accomplished. The student participants were certainly important members of their communities, but there was a desire to reach adult community members beyond the schools. We also wanted to have students interact more directly with community residents; have students develop and advocate for policies that would reduce concentrations of PM in their neighborhoods; and have a larger number of students involved in the project. These goals are important components of an expanded PM monitoring project the Alliance, the Watson Institute and organizational members of the Alliance hope to implement in the future. Since the Initiative, the Alliance and Watson Institute have conducted several small monitoring events in partnership with a non-profit organization and a community-oriented organization, but they were much less extensive because the resources were not available to hire a project manager for another large endeavor.

An Expanded PM Monitoring Project

The New Jersey Environmental Justice Alliance and John S. Watson Institute for Public Policy plan to implement another monitoring project that will expand the original New Jersey Urban Air Quality Education and Awareness Initiative by: 1) integrating the project more fully into the school curriculum, 2) involving more students, 3) reaching out to community members beyond the schools, and 4) performing more creative and purposeful monitoring.

Integrating the project more fully into the school curriculum and engaging more student participants

As previously indicated, Newark and Camden student participants in the first Initiative were members of science classes, while three of the four Trenton students were part of an environmentally-oriented mentoring program. After the monitoring, the Camden students prepared a PowerPoint presentation of their activities, and the Trenton students created both poster and oral presentations. An expanded monitoring project would be more fully integrated into school curricula by having: 1) science classes perform the monitoring; 2) math classes perform analyses on the collected data; and 3) social studies/civic classes develop, and advocate for, PM concentration reduction strategies, and conduct a community education campaign by making presentations to local organizations.

It is unclear whether integrating the project into the school curriculum in this manner would, in and of itself, increase the number of students involved in the project because it is not clear whether each class would be composed of different students, or if the same students would be in all three classes. This decision would depend on the educational and administrative needs of the individual schools participating in the project and would be the schools' decision. This is also true of the number of students in each class. While we would prefer to work with classes of no more than fifteen students, we understand that in an urban public school setting, classes often contain more than this number of students. The presence and active involvement of the students' teachers and proper project staffing should provide project organizers with the flexibility and capability to effectively involve a varying number of students in the project. However, the number of student participants would definitely increase because in this iteration of the project, the public high schools involved in the first Initiative would be paired with a faith-based high school from the same city. GreenFaith, a faith-based environmental advocacy organization, will be a partner in the expanded PM monitoring project and will identify faith-based high schools that wish to participate in the endeavor. Therefore, optimally we would have a science class, math class and social studies/civics class from two schools in a city involved in the project. The number of urban areas in which the project would operate simultaneously would depend on the amount of resources available to hire project staff.

Reaching the community and community empowerment leading to environmental justice

The environmental justice movement began largely as a grassroots movement (Bullard 1990; Gottlieb 1993) and now that it has become more institutionalized through the entrance into the environmental justice arena of mainstream institutions such as universities, environmental organizations, and governmental agencies, there is concern among environmental justice advocates that the grassroots portion of the movement will be subjugated to its more mainstream professional participants. This is a concern at both a national and local level and the Alliance has discussed this issue at length on several occasions. For this reason, the key addition to the expanded PM monitoring Initiative would be a more significant and focused attempt to empower student participants and other community residents.

We are confident that the expanded Initiative would empower an important part of the community, the student participants in the project, and provide them with the opportunity to empower other members of their community. An expanded monitoring project would individually empower student participants by imbuing them with knowledge and critical understanding of: 1) environmental justice; 2) PM air pollution in general; 3) the level of PM air pollution in their communities; and 4) health consequences of PM air pollution. As in the initial project, having a critical understanding of these topics may lead to an improved understanding by the students of societal factors that result in environmental injustices in their neighborhoods and also to a greater sense of personal obligation to their communities. Unlike the first Initiative, students participating in the second project would not only perform monitoring, but also use the background knowledge they gained, and the data they collected, to formulate and advocate for policies to reduce fine PM concentrations in their neighborhoods, and to educate other members of their communities regarding this pollutant. A large part of the advocacy efforts would most likely involve making presentations to local governmental agencies, state governmental agencies and quasi-private-governmental agencies on both a local and state level. By educating other community residents through presentations, developing PM reduction policies, and then advocating for these policies, the students will be engaging in activities that move them beyond individual empowerment and into empowering the community by participating in a process that is itself a form of community empowerment.

The community education performed by the students would be the key to individually empowering segments of the community outside of the schools. The students would make presentations regarding fine PM pollution in general and, more specifically, on the local data they collected, to as many community organizations as they could identify. The hope is that community residents educated, and therefore individually empowered, by these presentations would be inspired to formulate their own fine PM reduction strategies, or to at least pressure their local and state government officials to formulate and implement such strategies. We are also hoping the old adage that "knowledge is power" will apply, and that the knowledge and critical understanding of environmental justice and PM air pollution gained by residents from the students will move them to action and empowerment on a community level (Cargo and Mercer 2008).

Ultimately, we envision that advocacy efforts on the part of the students and other community residents will result in policy changes that improve the health of neighborhood residents by reducing local PM concentrations. This would be a much-desired example of a community empowerment outcome. Students and residents of affected communities advocating solutions to environmental justice issues based on their own ideas is the type of community empowerment that seems optimal to environmental justice advocates and the type the expanded Initiative would be intended to achieve. If it is conducted to its maximum potential, the expanded Initiative would serve not only as an educational and scientific project, it would also be a vehicle for intergenerational grassroots activities resulting in individual and community empowerment. Examples of previous community monitoring projects that achieved almost all of the goals discussed above are presented shortly.

More focused monitoring

If, as planned, more students participate in an expanded PM monitoring project, then the amount of monitoring can be increased. But even if the number of student participants does not increase, the monitoring performed would be more scientifically purposeful. The primary accomplishment of the monitoring in the first project was simply demonstrating that lay people could utilize the monitoring machinery and produce what we believe to be reliable data. The monitoring scheme was designed to reveal PM "hot spots" in city neighborhoods; areas where it was hypothesized that fine PM concentrations were relatively high. As mentioned earlier, there were, in fact, some locations monitored in the initial project that had concentrations which exceeded the 15.0 $\mu g/m^3$ federal annual standard for fine PM (40 CFR section 50.7(a)) even after adjustment using the calibration curve (Table 7.1).

However, the monitoring scheme can be refined to explore more subtle and interesting issues. For example, simultaneous monitoring could be performed in the downtown area of a city and in a nearby suburb in order to obtain a geographical comparison of PM concentrations. Similarly, simultaneous monitoring might be performed at a busy inner-city street corner that experienced a high level of diesel-powered traffic and a few blocks away in the "interior" of a neighborhood in order to determine the local extent of the influence of fine PM diesel emissions. These types of fine scaled monitoring are not being performed by NJDEP and could produce data that do not currently exist.

If the monitoring project could be institutionalized so that it occurred over a number of years, there is also the possibility of creating a database that would contain local fine PM concentrations. Using this database we may be able to assess a relationship between exposure to PM and local disease patterns, a major public health issue. The ultrafine PM concentrations

143

measured by the Ptrak would be of interest also since no ultrafine PM monitoring is currently occurring on a regular basis.

Previous Monitoring Projects

We are aware of several previous community-monitoring projects that can serve as models for the expanded monitoring project proposed in this chapter. Kinney et al. (2000) report on a community-monitoring project in which high school aged youth and professional researchers measured PM concentrations along the sidewalks of Harlem in New York City. The study was a result of an ongoing collaboration between two academic centers at Columbia University and WE ACT for Environmental Justice, a community-based environmental justice organization located in Harlem. Community residents eventually used data collected by the project to convince city officials to close a bus depot located near a local elementary school (O'Fallon and Dearry 2002).

Loh et al. (2002) discuss a project that monitored air pollutants, including PM, in the urban, mostly Of Color, neighborhood of Roxbury in Boston. The study was implemented primarily due to concerns about asthma initially raised by neighborhood youth working with Alternatives for Community and Environment, Inc. (ACE), a locally based environmental justice organization. Stationary monitors were installed in the neighborhood and data were made available to community residents by website, a telephone hotline system and a flag signaling system. The youth working with ACE were heavily involved in disseminating information about the project to the community and accomplished this in a variety of ways including presentations to various groups, and designing and implementing the aforementioned flag system that informed local residents about air quality on a daily basis. The data produced by the project were used to advocate for several policy changes such as converting diesel-powered buses to compressed natural gas and the relocation of a local bus yard. The project's youth participants were involved in this advocacy and directly communicated with policymakers during the course of their efforts. In addition to ACE, the collaboration was comprised of the Harvard School of Public Health, the Massachusetts Department of Environmental Protection, the North East States for Coordinated Air Use Management, and the Suffolk County Conservation District.

In a project that was at least partially inspired by the initial monitoring project conducted by the Alliance that was discussed earlier in this chapter, local high school students helped measure PM concentrations in Newark, New Jersey, in the spring of 2005 (Goldsmith and Gaddy 2006). The New Jersey Environmental Federation (NJEF) and CATF were partners in this project. The same monitors used in the Alliance's project were also utilized in this study and the study results were used by NJEF to help persuade several municipalities to pass resolutions against excessive idling by diesel-powered vehicles.

Although none of these previous endeavors contained all of the elements of the expanded monitoring project the Alliance wishes to implement, they can act as models for most aspects of an expanded project. Two of the projects included students directly in data collection, as did the initial Alliance monitoring project. But the Roxbury project also involved students to a much greater extent in the communication of information directly to community residents than the initial Alliance monitoring project did. This proves that the expanded monitoring project's extensive goals in this regard, i.e., student communication of project information to community residents, can actually be accomplished. Other goals of an expanded monitoring project that were fully or at least partially accomplished by all three previous projects were public policies developed, and then advocated for, by project participants. While policy development and advocacy occurred in all three projects, it is not clear from the literature whether youth project participants performed those tasks in Harlem and Newark. However, in Roxbury, it is clear that youth participants did both. Loh et al. (2002) report that, prior to the community-monitoring project, youth associated with ACE also developed and implemented strategies to reduce diesel emissions in Roxbury. In that instance, they held an "anti-idling day" and marched through neighborhood streets giving informational "pollution tickets" to drivers in an effort to educate the community about an anti-idling law and the negative health consequences associated with excessive idling. The Harlem project demonstrates that a monitoring project can play a role in community actions that change policy, since in that case monitoring data were used to help persuade city officials to move a bus depot.

It appears that the only aspect of an expanded monitoring project that these previous endeavors did not attempt to implement is an extensive integration of the project into an existing school curriculum. However, it should be noted that in Roxbury this may be because ACE already had an "in-school" curriculum that focused on helping youth identify and solve environmental health issues in their community. In addition, the aforementioned presentations made by youth participants of the Roxbury project included delivering a curriculum they developed to hundreds of students.

Taken together, these previous monitoring projects, and Roxbury related activities that occurred before the project, offer strong evidence that most, if not all, aspects of the expanded monitoring project proposed in this chapter are achievable.

General Characteristics of a Scientifically Oriented School-Based Project

The expanded PM monitoring initiative described in this chapter is a specific example of a scientifically oriented school-based project that furthers environmental and social justice goals. The general characteristics this type of project should have are discussed below.

The project should be a "hands-on" scientifically oriented endeavor that takes place at least partially in the community.

The hands-on aspect of the project is important because it distinguishes it from the normal classroom lecture format and thus makes it more interesting to the students. Having the project occur in the community distinguishes it from hands-on assignments performed inside the school such as laboratory exercises. The community location of the project should also make it more relevant to the every-day lives of the students. The monitoring project meets this criterion by having students measure PM concentrations at locations in their communities using handheld instruments. However, as with the initial monitoring Initiative, background information on the scientific, social and political aspects of the project can be provided to students during educational sessions in a classroom setting.

The project should address a substantive issue that is important to the community.

While having the project occur at locations in the community is an important aspect of making it relevant to the lives of student participants, even more critical is ensuring that it engages a substantive issue that is of importance to neighborhood residents. If the project concerns an issue that affects the well-being of the students and their families, they are far more likely to be truly engaged in the work than if they are learning about an abstract scientific concept that is difficult to link to their everyday reality, as may too often be the case in their science classes. Of course, the best way to ensure that the issue is important to the community is to have community residents, both students and non-students, participate in the selection of the topic in some manner. It is also important that the subject matter of the project be academically, intellectually and socially challenging. Picking up litter on city streets may be a worthwhile endeavor that can be performed in the community, but it is not ideal for the type of scientifically oriented, socially conscious project this chapter is advocating. The PM monitoring project is such a program because it addresses an issue that shortens the lives of urban residents in significant and disproportionate numbers, and presents students with interesting and complex scientific and societal challenges that help to develop critical thinking skills.

The project should allow students to formulate policies that address the issue.

Students should be given the opportunity to move beyond the scientific aspects of the project and create public policies that provide local solutions to the issue in question. Encouraging students to use their knowledge to formulate their own solutions to problems that affect them would be an important community empowerment experience. For the expanded monitoring project this would mean fashioning policies that would reduce fine PM concentrations in their communities. We understand that having students formulate policy will not be an easy task but there would be adults helping them, including their teachers and environmental justice advocates who are experienced at developing public policies. The youth interns at ACE are proof that urban youth can conquer this challenge.

The project should have the students educate other community members in some manner about the issue addressed by the project and about the project itself.

The project should have a community education component that is implemented by the student participants. This would be a community empowering activity and community residents would most likely be very receptive to hearing from the students because these young people would also be community members, and in many ways, would represent the hopes of communities that are suffering from a variety of ills. This is a critical component of any project because it would allow the project to reach community residents outside of the schools and hopefully lead to an empowerment of community members that is similar to the empowerment of the students. A community education campaign could take many different forms. In the case of the expanded monitoring project, student participants would make presentations to community groups on the work they have performed and on the dangers of fine PM.

If at all possible, the project should produce scientifically credible data.

A project should be pursued even if it does not produce scientifically credible data, but the collection of credible data would be extremely valuable. Reliable data should be packaged and formatted in a manner that makes it easily usable by students and other community residents and certainly would bolster the credibility of any policies that are formulated as a result of the project. The data could also be utilized by professional scientists who are performing research on the same or similar issues. Of course, if academic or other professional researchers are actually involved in the project, they may take steps to ensure that credible data are produced for professional reasons. In the case of the expanded monitoring project, producing scientifically credible data entails performing a calibration between the monitors used during the project and monitors used by NJDEP that are more accurate and precise.

If credible data are not produced, the type of project advocated by this chapter would still be worthwhile, especially to the community and to those participants who are not professional researchers. Information learned on environmental justice and on the issue in question should still result in individual empowerment and in community empowerment when it is transferred from students to community residents. This existing information may also be sufficient to allow polices to be created by project participants, particularly if other investigations have produced community level data pertaining to the issue. In fact, it might be wise to choose an issue that would allow the project to be successfully completed in all other aspects even if credible data are not produced. Implied in this discussion of data production, but perhaps it needs to be said explicitly, is that these scientifically oriented school-based projects are also research oriented.

Another important aspect of these projects, as with other research and non-research endeavors, is a project evaluation. The evaluation can be accomplished in several ways, for example, through pre- and post-project written surveys or in-person interviews. The method of project evaluation may legitimately vary but it is important that each project be subjected to some type of evaluation.

These general principles for a scientifically oriented school-based project are similar to the CBPR principles presented above, even though the two sets of principles were derived independently. We believe this is the case because both sets of principles are intended to guide research projects that will empower the community in some fashion. In fact, we view scientifically oriented school-based projects as a specific type of CBPR project. Therefore, the general rules we developed for these projects can be thought of as an application of CBPR rules to this specific type of project. We believe that if the generalized rules for scientifically oriented school-based projects are effectively implemented, then the intent of the CBPR principles will essentially be fulfilled.

After implementing the initial monitoring project we wanted to create an expanded monitoring project because we observed its potential to both educate students and empower the community. However, in addition to an expanded monitoring project, we also contemplated implementing other scientifically oriented school-based projects that focus on other environmental justice and health issues that are important to communities. For example, projects could focus on lead contamination in local housing or in soil, industrial contamination of the community's water table, whether contamination from a superfund site in a community has spread to adjacent locations, or determining whether a local body of water is clean enough for recreational use. We have attempted to use the lessons we learned from the initial monitoring project to plan an expanded monitoring project and to develop general principles for similar projects. We hope the general principles will help others who might wish to implement projects similar to the scientifically oriented school-based projects we discussed in this chapter.

To those creating such projects we suggest that the CBPR principles be used as guidelines while implementing the scientifically oriented school-based projects principles. If this sounds onerous, we hope fears will be allayed by reiterating that we believe effectively implementing the scientifically oriented school-based projects principles will essentially also be a successful application of the CBPR principles.

Increasing the Number of Scientific Projects with Links to Social Justice Issues

The overall theme of the book in which this chapter is included is using science to further social justice. There is no doubt that there are members of the scientific community who are concerned about social justice issues. Evidence of this is the attendance by hundreds of researchers at a 1994 symposium organized by the National Institute of Environmental Health Sciences on the research needed to ensure environmental justice (Shepard et al. 2002) and the recently formed, but relatively small, Environmental Justice and Science Initiative (EJSI 2011). The growth of participatory research might also be viewed as evidence of interest in social justice issues by academic researchers. However, many environmental justice (personal communication) and environmental (Reeves 2010) advocates feel that more scientific support for the environmental justice community, and more scientific projects with links to social justice issues, are needed.

We suggest that one way to sustain, and ultimately increase, the interest in the scientific community in conducting research that has implications for social justice is to increase the number of Blacks, Hispanics and Native Americans that receive doctorates from American universities in science and related fields. Although the numbers of Blacks, Hispanics and Native

Americans receiving doctorates in the sciences has increased over the past two decades, the percentages earned by these groups are still significantly below their representative percentages in the United States population. Native Americans and Alaskan Natives, Blacks, and Hispanics comprise 1.0%, 12.9% and 15.8% of the U. S. population, respectively (U.S. Census 2010). Yet in 2008, according to the Survey of Earned Doctorates (National Science Foundation 2009), American Indians, Blacks and Hispanics received only 0.4%, 4.5% and 5.5% , respectively, of the doctorates awarded in the life sciences by U. S. universities and 0.1%, 3.2% and 4.1%, respectively, of the doctorates awarded in the physical sciences.

We believe that since Blacks, Hispanics and Native Americans have faced racial injustice in the U.S. for centuries, scientists from these groups are more likely to perform research with links to social justice issues than their White counterparts. While we have no direct evidence to support this theory, there are data from a related field that allows an argument by analogy. Saha and Shipman (2008) found that underrepresented minority health professionals are more likely to provide health care to underserved populations than their non-minority counterparts, and this was true even if the non-minority health professionals were from lower economic backgrounds. The authors defined underrepresented minority health professionals as being African Americans, Mexican Americans, mainland Puerto Ricans, and American Indians/Alaska Natives, and underserved patient populations as disadvantaged racial minority groups (e.g., African Americans, Latinos and American Indians/Alaska Natives), Medicaid recipients, "the poor," the uninsured, and people living in federally designated Health Professional Shortage Areas or Medically Underserved Areas.

Saha and Shipman (2008) use their findings to argue in favor of programs that specifically target Of Color (racial minority) populations in an effort to increase the number of people Of Color graduating from U.S. institutions of higher learning with degrees in the health professions. We believe that the findings of Saha and Shipman (2008) regarding Of Color health professionals provides credibility to our argument that increasing the number of Black, Hispanic and Native American scientists is also likely to increase the number of scientific projects that have social justice implications. For that reason we urge the scientific community to increase their efforts to educate, graduate and employ more Of Color scientists.

Conclusion

There are numerous environmental justice issues that would be appropriate subjects for scientifically oriented school-based projects that could satisfy most, or all, of the general criteria presented above for these types of educational and research endeavors. We urge scientists, and members of the environmental and social justice communities, to collaboratively create, and implement, these types of projects that have the potential to help so many communities address environmental and other social justice issues.

References

Billig, S. H., 2000. Research on K-12 school-based service-learning, the evidence builds. Phi Delta Kappa, May 2000, pp. 658-664.

Butin, D. W., 2003. Of what use is it? Multiple conceptualizations of service learning within education. Teachers College Record, v. 105, pp. 1674-1692.

Brunekreef, B., A. H. Janssen, J. de Hartog, H. Harssema, M. Knape, and P. van Vliet, 1997. Air pollution from truck traffic and lung function in children living near motorways. Epidemiology, v. 8, pp. 298-303.

Bullard, R.D., 1990. Dumping in Dixie: Race, Class and Environmental Quality. Westview Press, Boulder, CO.

Bullard, R. D., P. Mohai, R. Saha and, B. Wright, 2007. Toxic Waste and Race at Twenty, 1987-2007, A Report Prepared for the United Church of Christ and Witness Ministries.

California Environmental Protection Agency, 2009. Methodology for Estimating Premature Deaths Associated with Long-term Exposures to Fine Airborne Particulate Matter in California. California Environmental Protection Agency, Air Resources Board, Draft Staff Report, December 7, 2009.

Cargo, M. and S. L. Mercer, 2008. The value and challenges of participatory research: Strengthening its practice. Annual Review of Public Health, v. 29, pp. 325-350.

Chung, A., P. Y. Chang, M. Kleeman, K. D. Perry, T. A. Cahill, D. Dutcher, E. M. McDougall, and K. Stroud, 2001. Comparison of real-time instruments used to monitor airborne particulate matter. Journal of the Air & Waste Management Association, v. 51, pp. 109-120.

Cook, W. K., 2008. Integrating research and action: A systematic review of community-based participatory research to address health disparities in environmental and occupational health in the USA. Journal of Epidemiology and Community Health, v. 62, pp. 668-676.

Dockery, D. W., C. A. Pope, X. Xiping, J. D. Spengler, J. H. Ware, M. E. Fay, B. G. Ferris, and F. E. Speizer, 1993. An association between air pollution and mortality in six U.S. cities. New England Journal of Medicine, v. 329, pp. 1753-59.

Environmental Justice and Science Initiative, 2011. Environmental Justice and Science Initiative. Available online at http://ejandscience.blogspot.com/.

EPA (U. S. Environmental Protection Agency), 2005. Review of the National Ambient Air Quality Standards for Particulate Matter: Policy Assessment of Scientific and Technical Information. Office of Air Quality Planning and Standards, December 2005.

GAO (United States Government Accounting Office), 1983. Siting Of Hazardous Waste Landfills and their Correlation with Racial and Economic Status of Surrounding Communities.

Godish, T., 1997, Air Quality, 3rd ed. Lewis Publisher, New York.

Gottlieb, R. 1993, Forcing the Spring. Island Press, Washington, D.C.

Goldsmith, A., and D. K. Gaddy, 2006. Diesel Hot Spots: A Snapshot of Newark, New Jersey. Finding a Path Towards "Kids Clean Air Zones." New Jersey Environmental Federation and Clean Water Fund.

Heaney, C. D., S. Wilson and O. R. Wilson, 2007. The West End Revitalization Association's community-owned and managed research model: Development, implementation, and action. Progress in Community Health Partnerships, v. 1.4, pp. 339-349.

Israel, B. A., B. Checkoway, A. Schulz, and M. Zimmerman, 1994. Health education and community empowerment: Conceptualizing and measuring perceptions of individual, organizational and community control. Health Education Quarterly, v. 21, pp. 149-170.

Israel, D. A., A. J. Schulz, E. A. Parker, and A. B. Becker, 1998. Review of community-based research: Assessing partnership approaches to improve public health. Annual Review of Public Health, v. 19, pp. 173-202.

Israel, B. A., E. Eng, A. Schulz, and E. A. Parker, 2005a. Introduction to methods in community-based participatory research for health. *In* Methods in Community-Based Participatory Research for Health, B.A. Israel, E. Eng, A. Schulz, and E. A. Parker (eds.). Jossey-Bass, San Francisco, CA, pp. 3-26.

Israel, B. A., E. A. Parker, Z. Rowe, A. Salvatore, M. Minkler, J. Lopez, A. Butz, A. Mosley, L. Costas, G. Lambert, P. A. Potito, B. Brenner, M. Rivera, H. Romero, B. Thompson, G. Coronado, and S. Halstead, 2005b. Community-based participatory research: Lessons learned from the Centers for Children's Environmental Health and Disease prevention research. Environmental Health Perspectives, v. 113, pp. 1463-1471.

Jerrett, M., R. T. Burnett, R. Ma, C. A. Pope, D. Krewski, K. B. Newbold, G. Thurston, Y. Shi, N. Finkelstein, E. E. Calle, and M. J. Thun, 2005. Spatial analysis of air pollution and mortality in Los Angeles. Epidemiology, v. 16, pp. 727-736.

Kinney, P. L., M. Aggarwal, M. E. Northridge, N. A. H. Janssen, and P. Shepard, 2000. Airborne concentrations of $PM_{2.5}$ and diesel exhaust particles on Harlem sidewalks: A community-based pilot study. Environmental Health Perspectives, v. 108, pp. 213-218.

Kittelson, D. B., 1998. Engines and nanoparticles: A review. Journal of Aerosol Science, v. 29, pp. 575-588.

Laverack, G. and N. Wallerstein, 2001. Measuring community empowerment: A fresh look at organizational domains. Health Promotion International, v. 16, pp. 179-185.

Levy, J. I., E. A. Houseman, J. D. Spengler, P. Loh, and L. Ryan, 2001. Fine particulate matter and polycyclic aromatic hydrocarbon concentration patterns in Roxbury, Massachusetts: A community-based GIS analysis. Environmental Health Perspectives, v. 109, pp. 341-347.

Loh, P., J. Sugarman-Brozan, S. Wiggins, D. Noiles, and C. Archibald, 2002. From asthma to airbeat: Community-driven monitoring of fine particles and black carbon in Roxbury, Massachusetts. Environmental Health Perspectives, v. 110 (Suppl. 2), pp. 297-301.

Massey, D. and N. Denton, 1993. American Apartheid: Segregation and the Making of the Underclass. Harvard University Press, Cambridge, MA.

Mohai, P. and B. Bryant, 1987. Toxic Wastes and Race in the U.S.: A National Report on the Racial and Socio-economic Characteristics of Communities with Hazardous Waste Sites. Commission for Racial Justice, United Church of Christ.

Mohai, P. and B. Bryant, 1992. Environmental racism: Reviewing the evidence. *In* Race and the Incidence of Environmental Hazards: A Time for Disclosure, B. Bryant and P. Mohai, (eds.). Westview Press, Boulder, CO.

National Science Foundation, 2009. Doctorate Recipients from U.S. Universities, Summary Report 2007-8, Survey of Earned Doctorates. National Science Foundation, Division of Science Resources Statistics, Directorate for Social, Behavioral, and Economic Sciences.

NJDEP (New Jersey Department of Environmental Protection), 2005a. 2005 Air Quality Report.

NJDEP (New Jersey Department of Environmental Protection), 2005b. Camden Waterfront South Air Toxics Pilot Project, Prepared by the New Jersey Department of Environmental Protection Division of Air Quality.

NJDEP (New Jersey Department of Environmental Protection), 2008. State Implementation Plan (SIP) for the Attainment and Maintenance of the Fine Particulate Matter ($PM_{2.5}$) National Ambient Air Quality Standard, $PM_{2.5}$ Attainment Demonstration Proposal, June 16, 2008.

O'Fallon, L.R. and A. Dearry, 2002. Community-based participatory research as a tool to advance environmental health science. Environmental Health Perspectives, v.110 (Suppl. 2), pp. 155-159.

Page, N. and C. E. Czuba, 1999. Empowerment: What is it? Journal of Extension, v. 37. Available online at http://www.joe.org/joe/1999october/comm1.php.

Pastor, M., R. D. Bullard, J. K. Boyce, A. Fothergill, R. Morello-Frosch, and B. Wright, 2007. In the Wake of the Storm: Environment, Disaster, and Race after Katrina. Russell Sage Foundation, New York.

Perkins, D. D. and M. A. Zimmerman, 1995. Empowerment theory, research, and application. American Journal of Community Psychology, v. 23, pp. 569-579.

Pope, C., R. T. Burnett, M. J. Thun., E. E. Calle, D. Krewski, K. Ito, and G. D. Thurston, 2002. Lung cancer, cardiopulmonary mortality, and long term exposure to fine particulate air pollution. Journal of the American Medical Association, v. 287, pp. 1132-114.

Pope, C., R. T. Burnett, G. D. Thurston, M. J. Thun., E. E. Calle, D. Krewski, and J. Godleski, 2004. Cardiovascular mortality and long-term exposure to particulate air pollution, epidemiological evidence of general pathophysiological pathways of disease. Circulation, v. 109, pp. 71-77.

Reeves, D., 2010. Scientists join activists in pushing co-pollution reductions in climate change bill. Inside EPA, June 18, 2010.

Saha, S. and S. A. Shipman, 2008. Race-neutral versus race-conscious workforce policy to improve access to care. Health Affairs, v. 27, pp. 234-245.

Schneider, C. G. and L. B. Hill, 2005. Diesel and Health in America: A Lingering Threat. Clean Air Task Force. Boston, MA.

Sheats, N., 2005. Preliminary Scientific Report for the New Jersey Urban Air Quality Education and Awareness Initiative, Prepared on behalf of the New Jersey Environmental Justice Alliance and the Center for the Urban Environment of the John S. Watson Institute for Public Policy of Thomas Edison State College.

Sheats, N., 2007. Preliminary Report on a Calibration of a TEOM and DustTrak, Performed for the New Jersey Urban Air Quality Education and Awareness Initiative, Prepared on behalf of the New Jersey Environmental Justice Alliance and the Center for the Urban Environment of the John S. Watson Institute for Public Policy of Thomas Edison State College.

Sheats, N., P. Montague and K. Jackson-Shrekgast, 2008. Comments on the New Jersey State Implementation Plan (SIP) Revision for the Attainment and Maintenance of the Fine Particulate Matter (PM$_{2.5}$) National Ambient Air Quality Standard, Submitted on behalf of the New Jersey Environmental Justice Alliance, Environmental Research Foundation Change to Win, International Brotherhood of Teamsters and Coalition for Healthy Ports.

Shepard, P., M. E. Northridge, S. Prakash, and G. Stover, 2002. Preface: Advancing environmental justice through community-based participatory research. Environmental Health Perspectives, v. 110 (Suppl. 2), pp.139-140.

TSI (Trust. Science. Innovation.), 2011. Particle Counters. Available online at http://www.tsi.com/en-1033/products/2136/p-trak_ultrafine_particle_counter.aspx.

United States Census, 2010. Available online at http://factfinder.census.gov/servlet/QTTable?-ds_name=PEP_2009_EST&-qr_name=PEP_2009_EST_DP1&-geo_id=01000US.

Ward, K., and L. Wolf-Wendel, 2000. Community-centered service learning moving from doing for to doing with. American Behavioral Scientist, v. 43, pp. 767-780.

Wilson, S., O. Wilson, C. Heaney, and J. Cooper, 2007. Use of EPA collaborative problem-solving model to obtain environmental justice in North Carolina. Progress in Community Health Partnerships, v. 4.1, pp. 327-337.

Yanosky, J. D., P. L. Williams, and D. L. MacIntosh, 2002. A comparison of two direct-reading aerosol monitors with the federal reference method for PM$_{2.5}$ in indoor air. Atmospheric Environment, v. 36, pp. 107-113.

Chapter 8: Between a Rock and a Green Place: Exploring the Relationship between Green Consumerism and Social Justice

Tendai Chitewere

Summary

In the past decade, green consumerism has helped to produce a new subculture of consumers who seek to balance a comfortable standard of living with environmental responsibility through the purchase of commodities deemed to be good for the environment. This green lifestyle focuses on emerging green technologies and products that allow individuals to make personal environmental choices in response to the global environmental crisis. One example of a green lifestyle is the creation of intentional communities, such as ecovillages, that attempt to create a sustainable way to live by building green homes, engaging in organic agriculture, sharing resources, and creating a sense of community. Based on an ethnographic field study of an ecovillage, this chapter aims to shed light on the complexity of creating a lifestyle that is environmentally sustainable and socially just. Given that green consumerism has been widely adopted, I raise the concern that the green lifestyle risks becoming a distraction to addressing current social and environmental injustices that disproportionately affect low-income neighborhoods and communities of color. An emphasis within the environmental movement on consumption of green consumer products excludes many people who cannot afford premium-priced "eco-goods." Moreover, if overconsumption is the fundamental problem with lifestyles in the United States, we need to ask whether a green lifestyle that is predicated on the consumption of a new category of commodities is the best way to confront environmental and social degradation. Ecovillages, if adopted for the urban setting, could provide an opportunity to create a lifestyle that focuses less on green consumption and more on socially and environmentally just ways to live sustainably.

Introduction

Current enthusiasm for all things green is hard to ignore. Unlike Kermit the Frog's melancholic song, "It's not easy being green," everything from nonprofit environmental organizations to corporate businesses are joyously proclaiming that it is not only easy being green, it's fun, fashionable and highly profitable (Brown 2001; Alsmadi 2007). An unrealistic embrace of an ill-defined "green economy" has generated hope that we can avoid an ecological catastrophe without interrupting our consumer oriented culture by simply buying the right kinds of material goods. Everything from solar panels to eco-vacations can now signal a consumer's commitment to environmental sustainability.

Increasingly, scholars and activists rightly argue that a lot of green consumerism is misleading and perhaps disingenuous. Greenwashing becomes a silkscreen for businesses to create a new way to market commodities in an age when consumption is itself under increased scrutiny (Durning 1992; O'Connor 1994; Price 1996; Smith 1998; Kaza 2000; Mander and Cavanagh

2007; Belli 2008). In green advertising, marketers are using their products to advocate a green lifestyle and "promote a corporate image of environmental responsibility" (Banejee et al. 1995). However, affluence and increased consumerism, green or not, is the cause of our environmental problems and, many argue, not the solution (Kaza 2000).

An emerging effort to address consumerism and environmental degradation is the development of ecovillages or ecological co-housing communities, which attempt to use the medium of community to 1) re-establish a sense of community and 2) use the community as a vehicle to increase sharing and thereby decrease the need for consumer goods. These efforts are commendable because they attempt to address the North American lifestyle, beginning with the way we live in our own homes. These new efforts reject neoliberal notions of private space and individualism and instead revert to the community as a source of social and environmental organizing, especially as it relates to proactive ways to respond to environmental and social degradation. Unfortunately, the neoliberal notions have been unsuccessful at addressing problems of equity. More specifically, critics of neoliberalism are particularly distressed by the focus that green consumerism places on the individual consumer rather than the collective citizen. And one of their major themes is that green consumption is inherently inequitable because, given the disparities between rich individuals and poor ones, equity is not a likely outcome of consumerism of whatever hue. Consumption, environmentalism, and equity have been linked before (Guha 2000; Kaza 2000), yet it continues to be sidelined in the public discourse. For example, representatives from southern countries present at the United Nations 1992 Earth Summit at Rio demanded that wealthy countries curb their disproportionate use of the world's resources (Kaza 2000). There are ample opportunities to examine the role of green consumerism and equity in the emerging Ecovillage movement.

Ecovillages, collective neighborhoods of residents who explicitly strive to live sustainably, are unusual. They are structured and intended to offer a collective approach to being green, while still emphasizing consumption, albeit green, as a way to minimize the ecological impact of their lifestyle. They have significant elements of collective decision making. For example, they have some of the elements of late 19th century ideas like Ebenezer Howard's Garden City, which aimed to bring people closer to nature by building sustainable communities away from the city they saw as unnatural (Howard 1902). Ecovillages also share some of the elements of a 1960s commune in that they strive to create shared governance that decentralizes power through consensus decision making and shared work groups (Durnbaugh 1997; Thies 2000). Simultaneously, they evince more modern trimmings of green consumerism in that residents explicitly own the inside of their homes as private property, while sharing the outside and other common amenities with the larger community. As such, they provide an excellent opportunity to probe what might be considered a good case of green consumerism, or at least one that tries to improve on the collective decision making, the lack of which so many critics of neoliberal reform cite as the chief problem. That is, while green consumerism is criticized as simply a misplaced response to environmental degradation, ecovillages demonstrates the complexity of proper consumption as a solution to environmental problems.

In this study, green consumerism, as expressed in ecovillages, is investigated as an effective means of implementing a commitment to the environment. Data for the case study of Ecovillage at Ithaca, a co-housing community in upstate New York, were collected as part of my doctoral dissertation. I lived in Ecovillage for approximately 15 months from 2000 to 2001, during which I engaged in participant observation and conducted informal interviews with residents in both the first resident group and the second neighborhood group. A third neighborhood (TREE) has been under construction and was scheduled to be completed in 2011. My ethnography provides an entry point for exploring the creation of a green lifestyle. The relationship between the environment, the community, and the consumption of green commodities can offer insight into the opportunities and challenges the green lifestyle has for ecological and social sustainability, and especially for addressing social justice. Ecovillages are well placed for such an exploration because they are intentionally designed to model sustainability; however, conspicuously missing from this effort is a discussion of justice as it relates to the environment.

Ecovillages offer a living laboratory in which to observe and analyze how green consumerism is deployed in an effort to create a way of life that does less harm to the environment than suburban sprawl and SUVs. The ecovillage project also offers a glimpse into the current limitations and future possibilities of creating a society that is socially just and environmentally sustainable. Through this project I raise the concern that the green lifestyle is increasingly distracted from environmental injustices that disproportionately affect low-income neighborhoods, communities of color, and communities with new immigrant populations. I argue this not because the green lifestyle is explicitly a social justice project, but because without addressing social injustice, we cannot reach social or environmental sustainability. Ecovillages, I suggest, need to and can contribute to just and sustainable community (Agyeman et al. 2003, 2005).

Green Places

The history of searching for a sense of community is long and varied; ecovillages come from a long lineage of efforts to find an idyllic way of life. Part of their lineage includes early utopian villages and communes, but it also includes the theory and practice of garden cities, as well as suburbanization, privatopias, and gated communities. The ecological co-housing community thus reflects on the mistakes of failed communities, adopts the best practices of the most successful, and thereby creates a promising design for modern living that attempts to address social and environmental degradation. I outline briefly the three community concepts that can be found in the co-housing movement and that are present in the ecovillage effort.

First, although they are certainly not 19th century utopias, ecovillages share some common traits with the ideas of utopias, specifically simplicity and harmony (Kanter 1972; More 2010). Utopian ideas have best been recalled as encouraging the transformation of societies into ones that are both socially and ecologically harmonious. The communal movements in the 19th century were varied in their focus; some sought religious freedom, some demonstrated communal work and a shared lifestyle by rejecting private property and labor, while others focused on egalitarianism, free love, vegetarianism and self-sufficiency (Thies 2000). Secondly, ecovillages include ideas of early 20th century city planners, Mumford, Geddes, and Howard, who advocated for the value of nature in the city. Howard's (1902) Garden City ideas can be found in Ecovillage at Ithaca through their effort to bring the everyday life of residents closer to the natural environment by incorporating an organic community supported agriculture (CSA) farm, a berry farm, and architecturally designing the houses to be surrounded by nature. Although critics of the garden city emphasize that suburban sprawl is the direct result of dismantling the city in favor of building in the country (Rodwin 1945), some ecovillages are built in cities and urban spaces.

Finally, because modern, sprawled forms of housing and neighborhoods in the United States have left households feeling isolated (Jackson 1985; Fishman 1987), ecovillages are responding to this disconnection by creating a community structure that requires social engagement and connections with neighbors. Ecovillages counter the design of suburban neighborhoods that center around automobiles and reinforce individualism through private driveways and sometimes gated neighborhood enclaves (McKenzie 1994; Caldeira 2000; Low 2001). Yet, walled enclaves share some common features with some ecovillages, such as their physical separation from the larger community and a cost of living that is higher than that of the surrounding neighborhoods, thus becoming exclusive, perhaps unintentionally.

Unlike gated communities, the ecovillage concept emerges from the intentional communities and co-housing movement in Europe. McCamant and Durrett's (1994) concept of co-housing in the United States focuses on the intentional physical design and social structure of clustering between four and thirty individually owned houses around a shared common house. In general, residents in co-housing participate in the design, governance, and the daily maintenance of their neighborhood. Ownership structures differ amongst the various communities, but residents typically own the inside of their home, while the external components such as the roof, siding, and yard are owned cooperatively by all households in the neighborhood. Individual households control their own financial resources and monthly maintenance fees pay for the general upkeep and insurance for the shared community property. A key design feature of co-housing is that vehicles are located outside of the neighborhoods, providing a pedestrian-friendly space for children to play safely between the houses and encourage spontaneous neighborhood interactions. The large windows that face into the common area allow residents to see each other and facilitate the development of social relationships. Shared community meals are also a key element in co-housing where a few nights a week, a volunteer cook-team prepares a meal for the neighborhood.

While most co-housing communities do not specifically identify themselves as ecologically focused, many informally adhere to principles that encourage resource conservation through sharing spaces and resources, such as gardening tools. Ecovillages are explicit about their endeavor to incorporate environmental sustainability into their practice of everyday life. According to the Global Ecovillage Network, ecovillages are "are urban or rural communities of people, who strive to integrate a supportive social environment with a low-impact way of life. To achieve this, they integrate various aspects of ecological design, permaculture, ecological building, green production, alternative energy, community building practices…" (GEN 2010). There are approximately eighty-five ecovillages located in the United States.

The location of nature is often viewed as separate from the built environment (Cronon 1996) and the place of being green is often equated with the nonhuman environment. The question of what is nature can sometimes be answered as being something antithetical to technology. So the tension between the countryside that embodies nature and the city, where the built environment can be characterized as outside of nature, has supported the development of green villages on the outskirts of the city. At the same time, there are a growing number of ecovillages that are urban; in fact, many co-housing communities in the Western United States are urban or semi-urban. Therefore, it is not surprising that green communities, like EVI, are often located in the suburbs near "nature." Although many ecovillages are connected to the city that surrounds them, being in nature helps residents to create their green identity. To protect this connection, conservation easements like land trusts are helping individual homeowners preserve their private land by providing legally binding rules that prevent future development. Land trusts and other efforts to protect open space are being transformed into symbols of sustainable community development.

Ecovillage at Ithaca (EVI) was formed in 1991 after a group of friends and family created and participated in Global Walk for a Livable Planet (Walker 2005). Participants walked across the United States with the mission of bringing attention to the unsustainable lifestyle of suburban sprawl and disconnected community. The grassroots effort that consisted of artists, teachers, college students, and professionals developed a strong sense of community through sharing meals with each other and with the residents of the communities they passed through. The walkers gave presentations and held workshops for the people they met along the way. They discussed the state of the environment in homes, churches and townhouses with anyone who was willing to listen. Some participants had to leave the walk early while new walkers joined along the way. One member who participated in the Global Walk recalled his experience as a truly grassroots movement. According to Jeff (names have been changed to protect the identity of the individual), a recurring theme as they passed through sprawling suburbs was that the people who lived there felt disconnected from their neighbors. Nostalgia for a sense of community was motivation to redesign the way homes and neighborhoods were built (Jackson 1985). The walkers discussed the need to have an alternative to suburbia. The leaders of the walk formed an EVI project interest group, which they would commence once some of the walkers had traversed Europe and Asia. Several of the members from the walk intended to participate in the ecovillage project, an idea they felt would be another grassroots movement merging social and environmental justice in a democratically constructed community.

Ecovillages offer innovative and exciting ways to use technology to address social and environmental degradation. If designing new homes, advocating for energy-efficient appliances, and sharing resources—while creating meaningful social bonds —is helping individuals make better choices, then the ecovillage projects offer a model that can possibly redesign the way we build new towns. However, if we are concerned about social justice and the opportunities ecovillages have to apply technology to address social injustice, ecovillages fall short. I suggest that one of the challenges facing the movement is a heavy reliance on green consumerism. Increasingly, green consumerism is dictating what it means to be sustainable, a concern echoed by many inside and outside the ecovillage movement.

Green Consumerism

Despite its prominence in our lexicon, green consumerism is not easily defined. It at once refers to the myriad of ways individuals use their consumer preferences to support and encourage environmentally benign products and services (Irvine 1989), while at the same time, suggests that green commodities are environmentally friendly, recyclable or made of recycled materials, biodegradable, etc. Green consumerism offers a "way for individuals to practice their environmental values, by purchasing from corporations that establish an eco-friendly brand identity" (Todd 2004). Yet, Smith (1998) suggests that "green consumerism is an act of faith" in that there is often little empirical evidence that commodities labeled as green are actually less environmentally harmful than their nongreen counterparts. She cautions that such ambiguity reveals green consumerism as a social myth, a form of greenwashing.

In a capitalist economy, one that depends on market activity to survive, it should come as no surprise that profit opportunities are increasingly driving mainstream discussions on environmental sustainability. The concern for nature and ecology that dominated the U.S. environmental movement in the 1960s has gradually been replaced by a focus on consuming our way out of the environmental crisis. Although unlikely bedfellows, environmentalism and consumerism are concepts that are increasingly finding their way into the fabric of everyday discourses on how to respond to the environmental crisis. Numerous scholars are rightly skeptical of this growing trend and argue that capitalism and environmentalism are antithetical precisely because of the central focus on consumption and consumerism (Mles 1993; O'Connor 1994; Sarkar 1999, Foster 2002). We know that consumption of natural resources has far exceeded the earth's ability to replace them, so it would seem that advocating for more consumption, albeit green, runs counter to the need to reduce our overall consumption of natural resources (Kaza 2000). In challenging us to consider the culture of consumption that dominates U.S. culture, Durning (1992) suggests that if consumption could be described as the hallmark of capitalism, then green consumerism needs to be scrutinized as ecologically problematic (Foster 2002). Scholars raise concerns that environmentalism is increasingly being drawn into profit driven directions (Jamison 2001). And often the result is greenwashing (Athanasiou 1996). Examples of early merging of nature and the economy might be seen in the Nature Company where nature, commodified such as natural bark and endangered species stuffed animals, could be bought in a built natural environment of a retail shop in a mall (Price 1996). While specialty businesses like Patagonia cater specifically to environmentally conscious consumers, a growing number of big box retailers are jumping on the green bandwagon (Smith, 1998). The problem arises when their green labeled commodities distract consumers from some of their environmentally damaging practices. Green marketing and the marketing of nature have emerged as alternatives to making sacrifices on consumption (Price 1996; Smith 1998; Chitewere 2006). The line between environmentalism and marketing is blurred with magazines like E! and Plenty that contain advertisements that mimic environmental education.

In this green consumerism, products are valued not just for their beauty and performance, but also on their supposed contribution to environmental sustainability. Furthermore, green marketing is creating a green aesthetic that functions as a form of personal identity and as a marker of class distinction (Todd 2004). A green lifestyle exists not so much by engaging in politically focused acts to address environmental degradation in a larger local and global social environment, as was the driving force in the 1970s environmentalism, but rather by focusing on a very personal expression of well-being. The challenge is that we risk reducing mainstream environmentalism to an exercise in conspicuous consumption (Luke 1998; Conca 2002).

Unfortunately, this focus on producing, marketing, and consuming green commodities is gaining traction in the United States and across the world (Hardner and Rice 2002; Alsmadi 2007). It is increasingly the preferred method of confronting a growing environmental crisis. While it may not be surprising that companies engage in green marketing in order to sell their wares or attract environmentally consciousness consumers, it is of concern when the reverse is true, when environmental activists advocate for green consumption as a solution to problems that have been caused by overconsumption. "Natural Capitalism" suggests that the principles of capitalism can be applied to ecological laws (Hawken et al. 1999). Proponents of green consumerism focus on shifting the burden of environmental conservation and protection onto corporate manufacturing and businesses.

Thus, what comes as an unfortunate surprise is the upscaling of environmentalism as it tries to keep up with the green joneses. This can be seen in the growth of eco-vacations, green consumerism, and perhaps the creation of a green lifestyle as it relates to ecovillages.

Green Consumerism and the Green Lifestyle

The green lifestyle model of EVI is significant to our understanding of consumption for three reasons. First, the green lifestyle suggests the importance of creating a sense of community as the first step to support environmental and social sustainability; through a strong cooperative community, neighbors are easily able to conserve resources through sharing and support each other in sustainable living. Second, the green lifestyle allows families to celebrate green technology and collaborate on purchasing energy-efficient commodities that might otherwise be too expensive for one family to afford. Third, focusing on the consumption of green commodities is expensive and time consuming, making them out of reach to people who do not have time or money. Residents often have long meetings to reach consensus on such decisions as whether paving the dirt road would be "natural." Intensive involvement by residents leaves very little time to address larger environmental and social problems that many communities struggle to resolve.

The consumption of green commodities is a primary means for residents to define their green lifestyle; it is also a source of internal conflict. When I asked individuals what made the community ecologically sustainable, many informants responded with consumables such as a hybrid cars, organic food, and sustainably harvested bamboo flooring. The personal environmentalism has been fed by the construction of nature as outside of human intervention and the aggressive marketing of being green (Castree and Braun 1998). One resident in the Second Neighborhood Group (SoNG) lamented that she felt privileged to be able to identify herself as living a green lifestyle because she could afford to buy a home in a green community. During the time of my research, there were some families who were not vegetarian or could not afford organic food and bought much of their food from one of the local chain food markets. Residents celebrated their ability to purchase environmentally friendly products, but one resident admitted that although she felt better about alternative choices, she knew substituting one commodity for another was not enough. Rather than emphasizing public transportation, hybrid cars were touted as greener than SUVs; instead of being critical of the two mile distance from Ithaca, some residents celebrated the short driving commute into town. The recent addition of a bus shelter is likely to increase the use of public transportation, but the quarter mile private road to the neighborhood still creates a physical barrier to the use of public transportation.

EVI, like many communities, is embedded within an unrelenting U.S. consumer culture; it is one of many responses to environmental and social degradation that fuel the drive to consume green commodities. Unlike the simple act of purchasing a single green product, ecovillages attempt to change the culture of how we live. Creating a green lifestyle means creating an identity through place and the consumption of green commodities, and thereby creating a means of explicitly identifying oneself as environmentally conscious. Many residents struggled to do something about the environmental crisis; the creation of a green lifestyle has been one way to answer that need.

Ecovillages are nestled within the context of contemporary U.S. environmental consciousness. New innovative energy-saving products and designs have been the focus of our collective efforts. The drive to green our lifestyle has reduced the emphasis on reducing our general consumption as a solution, despite the fact that consumption has long been identified as the leading environmental problem in the U.S. (Guha 2000). Green lifestyles by design focus on promoting alternative products to consume. EVI can be placed at the heart of this debate. On one hand it advocates for redesigning the human habitat by pioneering a sustainable culture (Walker 2005), while on the other hand, the community is illustrative of the tension between adopting green technology, identifying oneself as a citizen concerned with environmental degradation, and aspiring to live a lifestyle that is good for the environment, but which requires minimal personal sacrifice. Missing from this discussion is a concern for environmental injustice. Specifically, this kind of environmental engagement fails to respond to the problems that exist in environments and communities that are disproportionately burdened with the complex effects of environmental degradation. While important societal resources like improved transportation were frequently discussed within the community, these conversations, by the design of the project, rarely addressed the challenges of accessible public transportation in the wider community. Many residents within the community expressed frustration, and some moved away because they felt the structure of the project did not provide a mechanism to address environmental injustice. Thus I suggest that being green, as expressed in the green lifestyle of some residents, provided a proactive means to respond to environmental degradation, but at the same time, it provided a convenient way to avoid addressing environmental and social injustice.

When families move into an EVI neighborhood, they can immediately notice a reduction in their energy and water use. The green features of the community have been adopted to make it easy, convenient, and attractive to reduce some features of their ecological footprint (Trainer 1997). While the First Resident Group (FRoG) and Second Neighborhood Group (SoNG) used different design and construction models, both neighborhoods have similar basic features. In attempting to create a green lifestyle, EVI intentionally and communally incorporated simple green design features into each home.

All the houses in the EVI project were built as duplexes, allowing the houses to share a common wall that provides added insulation. Because poor insulation is a significant cause of energy loss, all the homes have double and sometimes triple paned windows. Passive solar design is maximized by the large south-facing windows that feature wooden trellises. During the hot summer months, the trellises are covered with kiwi and grape wines, protecting the inside of the house from the hot summer sun with a natural curtain of thick green foliage; in the winter months, when the sun is low and the ground is covered with snow, the crawling vines die back and allow the sun to heat the home. It is not uncommon for the temperature to reach the high 80s in the winter without turning on the thermostat. The houses in FRoG are all built vertically instead of horizontally in order to produce a small footprint. Five levels create separations between the kitchen/dining room and living room, bedrooms, internal loft, and bathrooms; this design uses less land and allows the FRoG to have thirty houses on two and a half acres. The homes themselves are small with the average two-bedroom, two-bathroom house measuring approximately 1500 square feet.

Other green design features were built into the FRoG neighborhood, but were not completed because they became too expensive for residents to implement. A gray-water recycling system was built into the neighborhood plumbing system; with a click of a switch, water from the bathtub, sink, and kitchen could be redirected to flush toilets and water gardens. Another design that residents often identified as a green feature, the "eco" in ecovillage, is the ability of the homes to accommodate solar panels that would take maximum advantage of the sun's rays. Many residents I interviewed were optimistic that eventually these green design features would be adopted as the demand went up and price came down, making them affordable. In the same spirit of making energy-saving easy and desirable, green appliances were chosen for the inside of the houses.

With one exception of a resident who wanted a gas oven, electric stove tops and convection ovens were pre-installed in the FRoG homes. Low-flow toilets and showers, and fluorescent light bulbs were just a few of the green, energy-efficient technologies that were designed into the homes. The residents were always looking for ways to improve their energy-efficiency; neighbors debated whether it was better to wash dishes by hand or in a dishwasher. The general consensus was that dishwashers were less green than washing dishes by hand and that the community laundry facility was designed to prevent individual households from needing their own laundry facility. At least one household owned a dishwasher and another owned a washing machine while I lived in FRoG, although they were generally not publicized. One resident, who later moved out, felt that she was not fully accepted in the community because she was not green enough; she did not recycle for example. Although she did not say that she was asked to leave, she often felt unwelcome. Her purchase and installation of a central air conditioner was visually unwelcome. Other evidence of the central role of green technology in the community was modeled by various green products used in the creation of the SoNG neighborhood.

Because residents of the SoNG built their neighborhood with individually hired architects and builders, the variety of green features was more extensive. Almost all of the houses are significantly larger than the ones in FRoG; the homes showcase innovative green design; two families chose to use straw bales for insulation. Several houses in the SoNG installed enough solar panels to turn their electric meter backwards, allowing them to sell electricity to New York State Electric and Gas during the sunny summer months and buy it back in winter. Several families installed composting toilets, passive solar design, and sustainably harvested bamboo for their flooring. The floors cover recycled plastic tubing that pumps hot water throughout the house and provides a unique heating system. The SoNG neighborhood installed an underground cistern that collects rainwater and snowmelt and stores it for use during the dry summer months. Thus the freedom of families to choose the kind of home they would build allowed them to select sophisticated green designs and appliances that were often more expensive, but were also identified and celebrated as essential for creating and modeling a green lifestyle.

When asked how their lifestyle is green, some residents responded that they drove a hybrid car, were members of the local CSA, and ate organic food. Other informants simply said that they lived in Ecovillage and therefore were doing a lot to be green simply by living in a small energy-efficient home. Being able to eat food that grows at EVI is one feature that almost all residents felt was an essential part of their green lifestyle. With an interest in the slow food movement, several residents are members of the CSA that is managed by two residents of the FRoG. Some residents celebrated the fact that they could live a very comfortable lifestyle without sacrificing comfort: the homes were attractive and modern; the surrounding land provides beautiful views of rolling hills and young forests.

The community helps residents identify with a lifestyle that is social and ecologically sustainable. Individual residents have access to the things that they feel help to support their commitment to the environment. For example, for Tina, being able to walk barefoot and meditate in the fields across from her home was very important. Living in Ecovillage allowed her, and many others, to feel physically and emotionally connected to each other and to the planet. Other practices reinforced this belief that they were living sustainably, such as when the volunteer cook-team would prepare an evening vegetarian meal using the crops that had been harvested that morning from the organic CSA.

Several couples admitted that they could have lived in a more traditional neighborhood but chose to downsize into EVI, in some cases going from 2500 square feet to 1300 square feet. Many, although not all families, purchased organic food and tried hard to buy locally. One resident owned chickens and sheep; she sold eggs to her neighbors and during a festive neighborhood event, invited the children to watch the shearing of the sheep. The bags of wool were shipped to California and returned as vibrantly colored balls of knitting wool that she made into hats and sweaters. Although some residents wished that the community were more ethnically and economically diverse, there were some opportunities for the community to talk about diversity as part of a sustainable community. According to the Communities Directory, there are currently 102 adults and 60 children living in EVI between the first and second neighborhoods; of those residents, ten percent identified themselves as a "person of color". Other features of the community that contribute to a green lifestyle include the prevalence of home schooling and unschooling of some of the children. During my stay, one child was unschooled, giving her the freedom to learn what and when she wanted. Although residents generally sought medical treatment when seriously ill, homeopathy and alternative healing were regularly included when confronting an illness. Regular meditation and yoga sessions were an ongoing part of the community.

Meditation, living lightly off the land and relying on cooperative sharing or unschooling are experiences that help define the green lifestyle. And while they do not necessarily require having access to a lot of money, they do suggest the need for time, space or land, and the cultural capital to participate. Who has access to this cultural capital and the doors it opens is unclear. What is clear is that ecovillages have not yet found a way to be inclusive of people outside of the green lifestyle (Chitewere and Taylor 2010). As upscaling of a sustainable way to live continues, the most vulnerable of our society are excluded. Furthermore, social justice as it relates to the environment in the form of environmental justice moves further into the margins as struggles to repair degraded environments are replaced with enjoying, appreciating and preserving beautiful nature. Scholars begin to doubt the effectiveness of a concept like sustainability as it comes to be used to mean almost anything to anyone (Harvey 1996; Chitewere 2006).

Environmental Justice and Green Lifestyles

Over the past forty years, the tension over how to define the environment and nature has made it frustrating for environmental justice scholars who have persistently pointed to the need for a broader definition of the environmental crisis that specifically exposes the poor and communities of color to greater risks from degraded environments (Bullard 1990; Bryant and Mohai 1991). Increasing interest, funding and support for environmental justice has encouraged EVI to include affordable housing and environmental justice concerns in village conversations and recruiting efforts. At the same time, an emphasis on green lifestyles, as manifested in the consumption of green technology, distracts well-meaning people from addressing environmental injustice.

The focus on green lifestyles as a meaningful way to create a sustainable community makes it easy to be green for some people, but not for many people who cannot afford to participate in the green economy as consumers. The focus of green lifestyles on consuming environmentally friendly commodities like hybrids or organic food means that less focus is given to responding to broader societal environmental solutions like public transportation and lead-free schools. Not only does the green lifestyle limit its critique of consumption itself as a fundamental problem in U.S. environmentalism (arguing for technological alternatives), it also limits who can participate in the dialogue.

By being inaccessible to low-income households and communities of color, green communities run the risk of contributing to a new kind of white flight—green flight. Instead of confronting polluters and working to improve spaces that are degraded, residents have chosen to move farther away from pollution and the most vulnerable people affected by it. In addition, the move onto former farmland is often in the same location as suburban sprawl, thus ecovillages risk creating green sprawl. Despite the real energy-saving features of these new communities, they stand as a sign of privilege and exclusivity that make their green lifestyle inaccessible to people who have continuously been marginalized and relegated to the worst social and ecological locations in our society.

This lack of emphasis on social justice was troublesome for many participants who eventually moved out of the community preferring to confront the ecological and social justice problems from within a different community model. Complicating this tension, the mission of the community was not clear to all residents or participants, especially for those who saw the project

165

as a grassroots social justice effort. During a discussion of the social justice aspects of EVI, an active resident remarked honestly and frankly that the neighborhood was not designed with social justice in mind, but rather as an alternative for upper-middle class families to live comfortably while consuming fewer resources. For this resident, and some others, the community would serve as a model for families who might otherwise build a "McMansion" on a hilltop. The innovative green features of the community and the access to organic food would make it an attractive and desirable alternative.

Although organic food was identified as a symbol of the green lifestyle, not all families who lived in the community could afford to consume organically. Some participants were afraid and embarrassed to admit this fact publicly. In a revealing and emotional experience, I offered to drive to the local store to purchase some tomatoes for a dinner a resident was preparing for her fellow neighbors; when I asked her whether I should purchase organic tomatoes, she became noticeably sad. I asked whether I had said something wrong and she responded that she could not afford to buy organically; I reassured her that most people in the United States could not afford to buy organic food either. While my response was comforting, the awkwardness of our brief exchange revealed a more complicated dynamic in the community, namely that the perception and desire to live organically and "be green" is often at odds with the reality that the green lifestyle is indeed not easy.

What is needed then is a more open and honest dialogue about the challenges and opportunities to create a lifestyle that is both socially just and ecologically sustainable. It will mean opening a discussion that often separates mainstream U.S. environmentalists from environmental justice activists. EVI has slowly begun to engage in this much needed conversation.

Ecovillage and Environmental Justice

Thus we see that the green lifestyle is trying to achieve balance between community and environmental conservation, and to achieve this important and worthy goal, residents are designing their community to foster sharing resources and cooperation. Through their location outside the city, architecture that promotes spontaneous social interactions, and through consuming green commodities, participants in the community believe they are able to realize their goal of social and environmental sustainability. But missing from this conversation is a response to environmental injustice. While we can celebrate the accomplishments of the project, questions of equity and justice can help explore the possibility of ecovillages as realistic solutions to our environmental crisis.

The EVI project is creating a sustainable village with four interrelated but separate entities. EVI has a 501(c)(3) nonprofit corporation (EVI, Inc.) that is charged with the mission of redesigning the human habitat by creating a model community and being an educational institution on sustainable living (Walker 1997). The initial goal of Ecovillage at Ithaca included the creation of five co-housing neighborhoods, a green village that incorporates more than one co-housing neighborhood. There are currently two completed neighborhoods, with the vision to construct three more neighborhoods that will constitute a village for approximately 500 permanent and temporary residents. The current director of EVI, Inc. is one of the project's founders who also helped to design the second co-housing neighborhood in the village. A board of directors advises the nonprofit, which functions as the public, educational arm of the project. Under the umbrella of EVI, Inc., the First Resident Group (FRoG), completed in 1996, is a separate private cooperative that is owned cooperatively by the residents. Residents of the FRoG hired a co-housing architect/design team to build all the homes in the neighborhood with the guidance of the residents. The FRoG bought their land from the EVI, Inc. nonprofit and work closely with EVI, Inc. to fulfill its mission. The SoNG was constructed five years later with future residents hiring their own architects and leasing their land from EVI, Inc. Together the two neighborhoods created a Village Association, which manages the jointly owned resources like the road and pond. Currently a third neighborhood (TREE) is forming. Based on McCamant and Durrett's (1994) co-housing model, both the FRoG and SoNG have homes that encircle a pedestrian-only walkway; the cars are located outside of both neighborhoods. In the FRoG, the homes are designed to be open, giving the residents a continuous view from the kitchen, dining and living room, and into the garden. The SoNG homes are, on average, larger than the homes in FRoG and many include more innovative green designs, like straw bale insulation, than the FRoG homes. A common house in each neighborhood contains a large community kitchen, dining space, laundry facilities, a guestroom, a children's playroom, and offices that residents rent. The common house provides resources that an individual household may not use regularly.

The site for the Ecovillage at Ithaca project on former farmland, just outside of the city of Ithaca, was selected because it would allow residents to feel physically, emotionally, and spiritually connected to nature and benefit from the preservation of the nature that surrounds the homes. At the same time, the proximity to the progressive college town of Ithaca was also appealing (Spayde 1997). The proximity to nature was part of what helped residents define what it meant to be "eco" in ecovillage, just as the adoption of a co-housing model helped residents articulate what it means to live as part of a socially sustainable village. The ecovillage project is part of a growing global trend to reconsider housing developments.

Technology plays an important role in EVI. Long, cold winters, rainy springs, hot summers, and a mild fall, describe the climate in upstate New York. With the variation in temperatures, residents harvest rainwater and retain solar energy through passive solar designs with south facing orientation and large double or triple paned windows. The inside of many homes are warm when the sun shines on cold snowy winter days. The neighborhoods are connected to the energy grid and the water and sewer lines of the City of Ithaca. The FRoG initially designed their homes to accommodate solar panels, but they were not installed because that green feature was too expensive. The SoNG has homes that harvest all or part of their electricity from solar panels. These design features help residents try to reduce their consumption of nonrenewable energy.

Although the EVI project has the mission to model sustainable social and ecological living, the everyday lives of residents in both neighborhoods are more complex. For example, the very location of the project was controversial; some families preferred to construct the community in an urban space, while others preferred being outside the city surrounded by nature. Some EVI residents and local community members were critical of the project on the grounds that it was contributing to sprawl. While some participants believed that the "eco" in ecovillage was best modeled by being physically located in the city where goods and services would be easily accessible through public transportation, biking or walking, others believed that it was more appropriate that a village centered on environmentalism be situated in nature. The location of the EVI project increased the cost of the project and prevented many of the participants from the Global Walk from joining.

The debates over the type of community EVI would become was divided along economic lines: those who saw the project as an alternative model of housing for wealthy families who could otherwise build a McMansion on private sprawling land, and those who envisioned a community that would model the tenets of economic, social, and environmental justice. The perspective that prevailed was one that would make the community unaffordable to many of the social justice participants who withdrew from the project. Despite the implicit decision to move away from addressing social injustice, many residents who stayed were still concerned about the exclusionary nature of the community and struggled to negotiate the tension of being a model community that would address social and environmental sustainability, while being accessible to only a narrow group of participants. Some residents believed that creating a comfortable community that could appeal to the wealthy was itself a worthy social and ecological achievement. A resident told me that it was the wealthy, with large homes on mountaintops and multiple gas guzzling cars, who were responsible for environmental problems. Thus, EVI provided an easy and accessible lifestyle option for those residents to reduce their consumption without compromising a comfortable lifestyle; a green lifestyle was achieved simply by living in the community.

Fulfilling the Promise of Ecovillages

Ecological co-housing communities are illustrative of a growing trend in U.S. environmentalism that I fear is moving away from addressing the public environmental crisis to focusing narrowly on personalized environmentalism; specifically, it focuses on impacts that affect the individual or a small group of people. This trend relies on green technology to model sustainable social and environmental living. But can we conserve the environment by consuming green commodities? If overconsumption is the fundamental problem with U.S. lifestyles, we need to ask whether creating a green lifestyle that is predicated on the consumption of green commodities is the best way to confront our current ecological and social problems. Ecovillages are uniquely placed to confront this dilemma and offer possible alternatives. Ecovillages can serve as models to reduce our consumption and dependence on consumerism, a problem identified as a primary challenge facing "overdeveloped" countries like the United States (Guha 2006).

One important social and environmentally sustainable aspect of the EVI project is that it attempts to create a culture of sharing. The common house, with its large kitchen and dining area, shared children's play space, community recreation room, laundry room, guestroom, and study and reading space all reduce the overall consumption of material goods in the community. By encouraging frequent, random interactions in the common house, on the land, and along the pathways, residents get to know each other in a way that makes it easy for them to share resources. It is common for a neighbor to borrow eggs, an iron, or a cup of milk when they are comfortable with each other. Residents can easily carpool into town and buy bulk organic food at a reduced rate. In addition to the emphasis on sharing goods, these new neighborhoods are engaged in the creation and maintenance of their community. Through consensus, many residents feel empowered to make valuable contributions to the governing of the place they live. Thus, perhaps the greenest feature of the community is not the eco-friendly technology; rather I suggest that it is the community's ability and willingness to share its resources amongst its neighbors. Instead of thirty washing machines for every home in the FRoG, there are three well designed, energy-efficient washing machines. (Although the community strongly discourages personal washing machines, while I was in the community at least one family did not adhere to this policy. They were frustrated with the inconvenience of using the common house machines and decided to purchase their own machine.) The CSA farm provides easy access to fresh fruits and vegetables as well as an opportunity for residents to be involved in the growing of their food.

The resources, governing and participatory activities that are designed into ecovillages are part of the fabric of many communities; apartment buildings, laundromats, city parks, and community gardens all contain the various aspects of the ecovillage movement that are accessible to a wider population and should be celebrated as being green. The primary difference is that participants in Ecovillage self-select; although the community is open to anyone who wants to join, the expense of purchasing a house in the neighborhood functions as a filter to people of lower incomes and, to a large extent, families who do not have the leisure time to participate in time-consuming consensus decision-making process.

Thus because of its emphasis on the environment as a place in nature, defining its green or eco-identity from commodities, and the strong desire to live a comfortable lifestyle, ecovillages risk being classified as an intentionally exclusive community not unlike the fortified enclaves of gated communities (Caldeira 2000; Low 2001).

Increasingly, community groups are creating ecological co-housing communities in urban low-income neighborhoods like Richmond, California (www.ecovillagefarm.org). Perhaps it is easy being green if we include the multiplicity of ways we approach the environmental crisis and seek viable solutions that truly take environmental and social justice into account and embrace the diversity of all our communities.

References

Agyeman, J., 2005. Sustainable Communities and the Challenge of Environmental Justice. New York University Press, New York.

Agyeman, J., R. D. Bullard, and B. Evans, 2003. Just Sustainabilities: Development in an Unequal World. The MIT Press, Cambridge, MA.

Alsmadi, S., 2007. Green marketing and the concern over the environment: Measuring environmental consciousness of Jordanian consumers. Journal of Promotion Management, v.13, pp. 339-361.

Athanasiou, T., 1996. The age of greenwashing. Capitalism, Nature, Socialism, v.7, pp.1-37.

Banejee, S., C. S. Gula, and E. Iyer, 1995. Shades of green: A multidimensional analysis of environmental advertising. Journal of Advertising, v. 24, pp. 21-31.

Belli, B., 2008. Green living deluxe: How the environmental movement went way, way upscale. E-Magazine. Available online at www.emagazine.com/archive/4417.

Brown, L. R., 2001. Eco-economy: Building an economy for the Earth. W.W. Norton & Company, New York.

Bryant, B. and P. Mohai P., 1991. Environmental Racism: Issues and Dilemmas. University of Michigan Press, Ann Arbor.

Bullard, R. D., 1990. Dumping in Dixie: Race, Class, and Environmental Quality. Westview Press, Boulder, CO.

Caldeira, T. P. R., 2000. City of Walls: Crime, Segregation, and Citizenship in Sao Paulo. University of California Press, Berkeley, CA.

Castree, N. and B. Braun,1998. Remaking Reality: Nature at the Millennium. Routledge, London and New York.

Chitewere, T., 2006. Constructing a Green Lifestyle: Consumption and Environmentalism in an Ecovillage. PhD. Dissertation, Binghamton University, Binghamton, NY.

Chitewere, T. and D. Taylor, 2010. Sustainable living and community building in Ecovillage at Ithaca: The challenges of incorporating social justice concerns into the practice of an ecological cohousing community. Research in Social Problems and Public Policy, v. 18, pp. 141-176.

Conca, K., 2002. Consumption and environment in a global economy. In Confronting Consumption, T. Princen, M. Maniates, and K. Conca (eds.). The MIT Press, Cambridge. MA, pp. 133-153.

Cronon, W., 1996. Uncommon Ground: Toward Reinventing Nature. W.W. Norton & Co, New York.

Durnbaugh, D., 1997. Communitarian societies in colonial America. In America's Communal Utopias, D. E. Pitzer (ed.). University of North Carolina Press, Chapel Hill, NC, pp. 14-36.

170

Durning, A., 1992. How Much is Enough? The Consumer Society and the Future of the Earth. W.W. Norton & Company, New York.

Fishman, R. 1987. Bourgeois Utopias: The Rise and Fall of Suburbia. Basic Books, New York.

Foster, J. B., 2002. Ecology against Capitalism. Monthly Review Press, New York.

GEN (Global Ecovillage Network), 2010. What is an Ecovillage? Available online at http://gen.ecovillage.org/ecovillages.html.

Guha, R., 2000. Environmentalism: A Global History. Longman, New York.

Guha, R., 2006. How Much Should a Person Consume? Environmentalism in India and the United States. University of California Press, Berkeley, CA.

Hardner, J. and R. Rice, 2002. Rethinking green consumerism. Scientific American, May 2002, pp. 89-95.

Harvey, D., 1996. Justice, Nature and the Geography of Difference. Blackwell, Oxford, UK.

Hawken, P., A. Lovins, and H. Lovins, 1999. Natural Capitalism: Creating the Next Industrial Revolution. Little, Brown, and Company, Boston, MA.

Howard, E., 1902. Garden Cities of Tomorrow. S. Sonnenschein and Co., London.

Irvine, S., 1989. Beyond Green Consumerism. Friends of the Earth, London.

Jackson, K. T., 1985. Crabgrass Frontier: The Suburbanization of the United States. Oxford University Press, New York.

Jamison, A., 2001. The Making of Green Knowledge: Environmental Politics and Cultural Transformation. Cambridge University Press, Cambridge, MA.

Kanter, R. M., 1972. Commitment and Community: Communes and Utopias in Sociological Perspective. Harvard University Press, Cambridge, MA.

Kaza, S., 2000. Overcoming the grip of consumerism. Buddhist-Christian Studies, v. 20, pp.23-42.

Low, S., 2001. The edge and the center: Gated communities and the discourse of urban fear. American Anthropologist, v. 103, pp.45-58.

Luke, T. W., 1998. The (un) wise (ab) use of nature: Environmentalism as globalized consumerism. Alternatives: Global, Local, Political, v. 23, pp. 175-212.

Mander, J. and J. Cavanagh, 2007. Beyond "Green Shopping." The Nation, September 4, 2007.

McCamant, K. and C. Durrett, 1994. Cohousing: A Contemporary Approach to Housing Ourselves. Ten Speed Press, Berkeley, CA.

McKenzie, E. 1994. Privatopia: Homeowner Associations and the Rise of Residential Private Government. Yale University Press, New Haven, CT.

Mies, M., 1993. Consumption patterns in the North: The cause of environmental destruction and poverty in the South. *In* Women and Children First: Environment, Poverty, and Sustainable Development, F. C. Steady (ed.). Schenkman Books, Rochester, NY, pp. 95-109.

More, T., G. M. Logan (ed.), and R. M. Adams (trans.), 2010. Utopia. W.W. Norton & Company, New York.

O'Connor, M., 1994. Is Capitalism Sustainable? Political Economy and the Politics of Ecology. The Guilford Press, New York.

Price, J., 1996. Looking for nature at the mall: A field guide to the Nature Company. *In* Uncommon Ground: Towards Reinventing Nature, W. Cronon (ed.). W.W. Norton & Company, New York, pp. 186-203.

Rodwin, L., 1945. Garden cities and the metropolis. The Journal of Land and Public Utility Economics, v. 21, pp. 268-281.

Sarkar, S., 1999. EcoSocialism or Eco-Capitalism? A Critical Analysis of Humanity's Fundamental Choices. Zed Books, London.

Smith, T. M., 1998. The Myth of Green Marketing: Tending our Goats at the Edge of Apocalypse. University of Toronto Press, Toronto, ON.

Spayde, J., 1997. Ithaca, New York: Our kind of town: A gritty upstate city where the grassroots are green. Utne Reader. Available online at http://www.utne.com/Politics/Americas-Most-Enlightened-Towns-Ithaca-New-York.aspx.

Thies, C. F., 2000. The success of American communes. Southern Economic Journal, v. 67, pp. 186-199.

Todd, A. M., 2004. The aesthetic turn in green marketing: Environmental consumer ethics of natural personal care products. Ethics & The Environment, v. 9, pp. 86-102.

Trainer, T., 1997. The global sustainability crisis: Implications for community. International Journal of Social Economics, v. 24, pp. 1219-1240.

Walker, L., 1997. EcoVillage at Ithaca: Phoenix rising. CoHousing: Contemporary Approaches to Housing Ourselves, v. 9, pp. 8-10.

Walker, L., 2005. Ecovillage at Ithaca: Pioneering a sustainable culture. New Society Press, Gabriola Island, BC.

Part III: Rediscovering the Commons: Philosophical Principles for Global Environmental Justice

Chapter 9: Buddhist Living in the Anthropocene

Jill S. Schneiderman

Summary

Because of the extended time frame over which they occur, human-induced environmental changes are out of sync with human lives lived in an age characterized by nanosecond attention spans. As a result, the violence exacted by such changes poses representational and motivational challenges to human abilities to address them. I tackle the question, as an earth scientist, obviously from outside the discipline of religious studies: How might Buddhist thought provide valuable tools for people interested in working progressively at the intersection of violence and human-induced environmental degradation? Like other non-specialist Westerners, I have cobbled together an eclectic Buddhist perspective. In this chapter I explore affinities between certain Buddhist themes and my own professional orientation and expertise in earth science. To that end, I describe concepts of time derived from geological science and Asian mythical traditions; probe ideas about violence in relation to global environmental degradation; and, consider implications of the Noble Eightfold Path, the way to deliverance from suffering taught by the Buddha, for adjusting systemically to environmental change over time. By expanding, scientifically and spiritually, what the present moment can be, an instant of infinite duration, I hope to connect my journey towards taking a long view of a moment with the long view of change that human beings require to awaken to violence that is at once potentially catastrophic and so slow that it's difficult to discern and therefore counter.

Introduction

"'Tell me, children,' I would begin, 'what do our old myths have in common with geology?'

'…Goddesses, children,' I would announce in triumph. 'Don't you see? Goddesses are what they have in common.'

',.. Think about it,' I would say, 'and you'll see. It's not just the goddesses —there's a lot more in common between myth and geology. Look at the size of their heroes, how immense they are—heavenly deities on the one hand, and on the other the titanic stirrings of the earth itself—both equally otherworldly, equally remote from us….And then, of course, there is the scale of time—yugas and epochs, Kaliyuga and the Quaternary. And yet—mind this!—in both, these vast durations are telescoped in such a way as to permit the telling of a story.'

...And, to follow this, I decided, I'd tell them the story of the Greek goddess who was the Ganga's mother. I would take them back to the deep, deep time of geology…."

Amitav Ghosh, 2005. The Hungry Tide, pp. 150-151.

Because of the extended time frame over which they occur, human-induced environmental changes—atmospheric ones such as increased temperature, rising sea level, and unpredictable storm patterns, as well as those that are land-based, including desertification and drought—are out of sync with human lives lived in an age characterized by nanosecond attention spans. As a result, the violence exacted on all living beings by such changes poses representational and motivational challenges to human abilities to address such ills. In this chapter I tackle the question: How might Buddhist thought provide valuable tools for those of us interested in working progressively at the intersection of violence and human-induced environmental degradation? To that end, in this chapter I describe concepts of time derived from Asian mythical traditions and geological science; probe ideas about violence in relation to global environmental degradation; and, consider implications of the Noble Eightfold Path, the way to deliverance from suffering taught by the Buddha, for adjusting systemically to environmental change over time. By expanding, scientifically and spiritually, what the present moment can be, an instant of infinite duration, I hope to connect my journey towards taking a long view of a moment with the long view of change that human beings require to awaken to violence that is at once potentially catastrophic and so slow that it's difficult to discern and therefore counter.

The Realm of the Eternal Moment

From perches that usually encompass great swaths of space, we geologists view varied earth processes that include changes of landscapes over time. As a result, our perspective also takes in vast sweeps of time. I'd say we imbibe deep time. Naturally, we discern clues hinting at the qualities of systems that operated in bygone eras. In outcrops of rocks, forgotten fossils, and minute mineral fragments, we find evidence of earlier events on earth. And to us, present time mingles intimately with the past and the future. Ours is a cultivated skill that requires patience grown from sitting still or walking slowly in the field, and watching nothing happen. Most other scientists derive understanding by observing processes occur. By virtue of our mode of seeking knowledge, geoscientists have much in common with Buddhist thinkers. Though Buddhist approaches to time focus on the present and geological thinking melds temporalities, together they hold promise for a compassionate, time-transcendent vision required for what Stephanie Kaza (2008), author of Mindfully Green, calls a green practice path.

Sounding as if he might be describing a geologist, in his work Images and Symbols, the great historian of religion Mircea Eliade (1952) describes the intermediate situation of man who, although living in his own historic time, keeps avenues open to what he calls Great Time, and never loses consciousness of the unreality of historic time. Eliade retells the Indian myth of Indra, a King of the Gods, who, visited by Vishnu in the form of a beggar boy, studies a parade of ants in the great hall of the palace and in doing so is cured of his pride and ignorance. According to Eliade, Indra achieves humility by transcending mundane time and entering into mythical and cosmic Great Time. In a sequel to the myth, Indra, humbled and withdrawn into

174

ascetic life, listens to spiritual guide Brihaspati speak about finding fulfillment in the current-day world by uniting contemplation with action. As a result, Indra comes to understand the importance of not abandoning one's historical situation while searching for a universal Being, but rather engaging compassionately the responsibilities of the historical moment while keeping in mind the perspectives of Great Time.

Eliade also describes the symbolism of the seven steps of the Buddha at the time of his birth as the Buddha transcending space and time. He writes of how in both Hindu and Buddhist traditions, time and timelessness, or time and eternity, are two aspects of the same principle, *khana*, a Pali word meaning the present moment. Eliade explains that *khana* also means opportunity and that, according to the Buddha, one whose thought is stable lives under the admirable condition of an eternal present in which time no longer flows. In the Buddhist scripture of <u>Connected Discourses</u> (Bodhi 2002), the Buddha congratulates monks who have "seized their moment." Eliade characterizes this favorable moment as a paradoxical instant that suspends duration; he interprets it as ecstatic enlightenment that is mystical and beyond time.

That time and timelessness can lose their tension as opposites resonates in the experiences of most geologists. In some ways, geologists experience the flow of time differently than most other people. In reading the record of rocks, we geologists read the rhythms of cosmic Great Time. We live in circumstances that can be compared to Eliade's description of the Indian mode of experiencing time. Other thinkers, some of them geologists, have also written of this. Isabel Fothergill Smith, an early 20th century geologist and educator who at age 97 befriended and kept company with me until she left this world just shy of her 100th birthday, is remembered as having set an intellectual agenda for her Scripps College students that would encourage what she called "the wide view." An interview transcript among her papers at Scripps College reveals her perspective (Smith 1971): "Geology…is the one science that makes you see the world whole, as no other science does…Geology is simply physics on an earth scale and chemistry on an earth scale and biology from the fossil pattern on a world scale, looking back in time. Really, if you study geology, you do see science whole and you do see the world in both ways, horizontally and vertically back in history."

Smith's (1956) whimsical musing about a field trip with her students vivifies the timeless and time-full realm of the geological: "One need not be a trained geologist to learn something of earth's story and history. What we could read in the rock structure around us on the desert revealed that this bit of landscape had been eons in the making, had known both quiet and turbulent times. We were not the first to find fossil fragments in the rock beds; geologists had reported the finding of bone fragments of a primitive horse, of camels, antelopes, even bears and a mastodon and the shell of a fresh-water mollusk, all of which spoke of a climate in the long past far less arid than now, an open country congenial to grazing mammals, with here and there good water holes, and nutritive grasses where today only the most drought-resistant shrubs can endure. We knew there had been periodic volcanic activity in the region for among the rock layers were thin beds of white volcanic ash; and saline deposits were a clue to intervals of little rain and excessive evaporation of the lake waters…The rock layers proved to be a rich record of history." Her writing demonstrates the fluid ease with which some geologists' bodies and minds move through space and time.

How do we do this? Using Eliade's (1952) phrase, I would describe geologists' ability as the power to "return backwards." Or, as Pablo Neruda wrote and Alastair Reid translated in "Keeping Quiet," (Neruda and Reid 2001), we let the earth teach us:

"If we were not so single-minded
about keeping our lives moving,
and for once could do nothing,
perhaps a huge silence
might interrupt this sadness
of never understanding ourselves
and of threatening ourselves with death.
Perhaps the earth can teach us
As when everything seems dead
And later proves to be alive."

For example, this year I've been living on a tiny coral island in the Atlantic Ocean. Here in Barbados, everywhere I look with my geologist's gaze I see evidence of changed landscapes and past climates. A coral island that rose roughly 1200 feet above sea level in the last one million years, Barbados is a geological infant. Though some sandstone and shale form a nucleus of the island, more than 85% of the exposed land consists of coral rock, known to geologists as limestone, naturally lithified from broken debris of ancient coral reefs. The island is unique in the Caribbean. Unlike the Bahamas that consist largely of windblown sand cemented together by the action of rainwater, or other Caribbean islands so vividly volcanic, Barbados today is comprised of nothing more than subaerial coralline remnants of dead communities and submarine fringes of currently living colonies of organisms—corals. Each tiny individual coral animal, called a polyp, which is related to and looks like a sea anemone, encloses itself in a stony cup of limestone that it secretes. As they grow, the polyps divide to form coral colonies that build up on top of each other and manifest as a reef. Over thousands of years, coral reefs respond to fluctuations in sea level as well as changes in water temperature and other environmental conditions.

While dwelling on this island, I traverse slopes telling me that where I now walk, ocean waves once lapped. Hillsides shaped like treads and risers of a coralline staircase, coastal terraces in geological parlance, mark ancient shorelines. These old coastal features some distance above the modern coastline indicate that with tectonic uplift, changed climate, and sea level fluctuations, some colonial organisms have become extinct while others have succeeded them. I think of these ancient reefs as paleo-Sanghas, extinct communities that lived and died together. (*Sangha* is a

176

Buddhist term in the Pali language referring to the assembly of all beings.) From the tops of the coral staircases, I stare down through clear, aquamarine waters to the living coral communities and can anticipate their end, brought on eventually by tectonic shifts, rising temperature, or nonpoint-source pollution.

Other sources of Buddhist thinking about the idea of the eternal present resonate with my experience as a geologist of the flow of time. For example, Eihei Dogen , the 13th century Japanese thinker who is remembered as the founder of the Soto School of Zen Buddhism, devoted "The Time-Being," an important philosophical fascicle of his Treasury of the True Dharma Eye, to a discussion of his recognition that "time itself is being, and all being is time" (Dogen and Tanahashi 1995). For Dogen, time consists not of the past, present and future so much as it does of events, moments and movements. In "The Time-Being," Dogen admonishes his students to: "See each thing in this entire world as a moment of time…Do not think that time merely flies away…In essence, all things in the entire world are linked with one another as moments." As a geologist, I see some fundamental truth in what Dogen teaches for, although time has a linear direction and we couldn't have evolution without it, the cyclicity of earth processes, like the growth, death, and regrowth of Barbados' coral reefs, speaks to the timelessness of moments as depicted by Dogen. Lastly, in his "Mountains and Waters Sutra," Dogen speaks as if he were himself a geologist: "Now ordinary fools and mediocre people think that water is always in rivers or oceans, but this is not so. Rivers and oceans exist in water. Accordingly, even where there is not a river or an ocean, there is water. It is just that when water falls down to the ground, it manifests the characteristics of rivers and oceans…The existence of water is not concerned with past, future, present, or the phenomenal world." What we have here is a 13th century description of reservoirs of the hydrosphere and acknowledgement that the parts make up an endlessly cycling and timeless whole. It is in the realm of eternal moment that the thinking of geologists and Buddhists overlaps.

Violence and Environmental Degradation

In attempting to address the question of how humans should behave as actors in the system of environmental change, I find particularly useful Rob Nixon's (2006-2007) concept of slow violence. In his remarkable paper, "Slow Violence, Gender and Environmentalism of the Poor," Nixon writes evocatively of what he calls the oxymoronic notion of slow violence, or acts whose "lethal repercussions sprawl across space and time." Just as Stephanie Kaza (2005) celebrates marine scientist Rachel Carson's contributions to the domain of thinking about time, Nixon also raises Carson's ghost, quoting her description in Silent Spring of the cost of toxic pollution as "death by indirection" (Carson et al. 2002). Nixon employs vivid words to help shape an image of slow violence: oblique, slow acting, non-instantaneous, unspectacular, nondramatic, and amorphous. He describes the results of slow violence as "attritional calamities" with "deferred consequences and casualties"; "dispersed repercussions" that "pose formidable imaginative difficulties" because they "star nobody." He cites as examples of such "convoluted cataclysms" desertification and deforestation and, like Kaza, he identifies accumulating greenhouse gases and consequent climate change as most ominous. For Nixon, slow violence is synonymous with global environmental degradation and we must call attention to these changes so that we can right the ills associated with them. But, he asks, how can we enliven catastrophes that are "low in instant spectacle" but "high in long-term effects"? Since they pose overwhelming representational challenges, we must summon creativity he says. I would add that global environmental changes that are out of sync with human lifetimes—Nixon's "environmental time"—are not only difficult to represent. These catastrophes present motivational challenges as well. We must render slow violence *actionable* as well as visible.

How can Buddhist philosophy be applied to the problem of slow violence? Famously non-violent and, as we have seen, at peace with the notion of all time being present in the present, Buddhism offers a helpful approach. However, in order to see this, we must first understand slow violence in terms of the essential typology of violence proposed by Norwegian peace scholar Johan Galtung (1969) in his foundational paper "Violence, Peace and Peace Research." As Galtung points out, traditionally, violence had been thought about as personal violence only, with important subdivisions in terms of violence vs. the threat of violence; physical vs. psychological; intended vs. unintended. His central point in the paper, however, is to call attention to and deem most basic, the distinction between personal and structural violence. He also refers to personal and structural violence as direct and indirect violence, respectively. The primary distinction for Galtung, between personal and structural violence, is that, in the former, there is a concrete agent, whereas in the latter, there is no such actor. Instead, violence is built into the structure of a system and shows up as unequal power and unequal life chances. Although for Galtung, there are many dimensions of violence, by his definition, violence is the cause of the difference between "what could have been and what is." By way of example, Galtung states that in a society where life expectancy is twice as high in the upper as in the lower classes, violence is exercised even if one can point to no concrete actors that attack others, for example one person killing another. This description of structural violence has obvious corollaries in, for example, children of the lower classes who die of diarrhea because their drinking water quality is terribly degraded. Such environmental degradation can then be thought of as structural violence by Galtung's definition.

Despite the similarity between Galtung's structural violence and Nixon's slow violence, a subtle but important distinction remains to be made between them. Structural violence conveys stasis whereas slow violence acknowledges movement and hence the possibility of change. Galtung clearly intended to convey stasis; he says that structural violence shows stability over time when compared to personal violence, which displays substantial temporal fluctuations. Describing what might be thought of as the inverse of Nixon's amorphous causalities of slow violence, Galtung writes: "Personal violence *shows*. The object of personal violence perceives the violence, usually, and may complain—the object of structural violence may be persuaded not to perceive this at all. Personal violence represents change and dynamism—not only ripples on waves, but waves on otherwise tranquil waters. Structural violence is silent, it does not show—it is essentially static, it *is* the tranquil waters...In a *static* society, personal violence will be registered, whereas structural violence may be seen as about as natural as the air around us."

But as Dogen recognized centuries ago, even tranquil waters move, albeit sometimes imperceptibly, and in the four decades that have elapsed since Galtung introduced his term, we have come to live time very differently. By using the term *slow* violence, Nixon highlights how our experience of the flow of time has changed. His concept sparks engagement with Buddhist notions of the eternal moment and impermanence. This is important because in an age of instant connectivity and contracted attention spans, it is now even more difficult to draw attention to the violence of environmental degradation.

Indeed this is the case with our present crisis in which global climate change is rendered natural in the face of increased greenhouse gas production, considered necessary for industrialized society. When he published his work in 1969, Galtung observed that the social structure that gives rise to such slow violence is itself sluggish and may not be changed quickly. Thus we can see that Nixon's particular form of slow violence in the form of creeping environmental degradation is a type of structural violence that features the juxtaposition of slow change in earth systems with the sluggish nature of social systems. Importantly, Galtung (1990) later expanded his conceptualization of violence into a third realm that he called cultural violence—violence that obscures and/or legitimizes personal and structural violence and helps to promote a culture of impunity among its perpetrators. Because he found the idea useful but the phrase problematic, political geographer Joe Nevins (2002) adapted Galtung's idea, calling it symbolic violence. The idea that violence can be simultaneously structural, slow and symbolic pertains significantly today when environmentally challenging situations are merely discussed rather than acted upon attentively. Human time flows in brisk countercurrent to slow-motion environmental time, and guarded agents in lethargic systems listlessly search for avenues of action, knowing not in what direction to move. But to my mind, the Buddha's Noble Eightfold Path offers multiple approaches to countering the slow violence of climate change and other forms of environmental degradation that masquerade as inevitable and perhaps benign forms of global environmental change, Galtung's (1969) "tranquil waters."

Nonviolence: Wise-Hearted and Kind-Minded Beings not Hard-Hearted Intellects or Good-Hearted Fools

The Noble Eightfold Path is the last of the Buddha's Four Noble Truths. It is considered to be the Way leading to the end of suffering (*Dukkha*). Comprised of eight divisions, the entire teaching of the Buddha deals in some way or other with this Path. The eight categories are intended to promote the three essentials of Buddhist training: wisdom (*panna*), ethical conduct (*sila*), and mental discipline (*samadhi*):

Wisdom

1. Right Understanding

2. Right Thought

Ethical Conduct

3. Right Speech
4. Right Action
5. Right Livelihood

Mental Discipline

6. Right Effort
7. Right Mindfulness
8. Right Concentration

Undertaken more or less simultaneously, they facilitate development of peaceful individuals and harmonious societies (Rahula 1959). Although linked together and mutually reinforcing, to gain a clearer understanding of the eight divisions of the Path, it helps to group them into these three categories. Wisdom represents the qualities of the mind such as intellect. Ethical conduct arises from qualities of the heart, the love and compassion for all living beings on which the Buddha's teaching is based. According to Buddhist principles, a perfect person develops these two qualities equally. A Buddhist life should inseparably intertwine wisdom and compassion. Given the pairing of wisdom-head and ethical conduct-heart, mental discipline in some sense can be linked to the entire body in the form of varied contemplative practices.

Much has been written about, and contemporary mindfulness communities have focused on, development of steps of the Noble Eightfold Path pertaining to mental discipline. This is not surprising because these steps involve practices that individuals can undertake separately and then experience directly the positive effects on the mind and body. While many practitioners have begun different types of contemplative practices (e.g. stillness, movement, relational) as an individual opportunity to enhance peacefulness in themselves and on the planet, on the other hand, scholars predictably have worked to articulate knowledge gleaned about the earth to help society cope with problems arising from human-induced global changes. Indeed, since the early 1980s scientists have called attention to global climate change, warning us that humans have influenced to ill effect the chemical and physical composition of the atmosphere and hydrosphere. As those who have heeded these pronouncements know, we have squandered precious time by debating the veracity of phenomena that are clear to those awake enough to see global change in the form of rising sea level, melting glaciers and disappearing sea ice, to name just three visually stunning examples.

What has been absent from the conversations, or at least only marginally heard, is a focus on the ethical conduct component of the Eightfold Path in relation to global environmental degradation (Stanley et al. 2009). Given that scientists in particular have been heard most loudly on the subject and given the proudly professed divorce of head from heart in the scientific enterprise, I am not shocked that ethical conduct motivated by love and compassion has not been at the forefront of the global change conversation. It is exactly this fundamental pillar of Buddhism—heart—that we need in order to attend to slow violence as described by Nixon. To counter the violence of environmental degradation and its effects on all living beings, leaders in the global change arena, including scientists and faith communities, must hold themselves to the specific exacting steps outlined in the Noble Eightfold Path. For example, climate change negotiators must hold themselves to right speech or they will end up engaging in representational violence, in this case speech that obscures the slow violence of climate change.

Returning to the central challenge of slow violence and the need for representation, imagination and motivation, human society today needs startling icons to vivify deleterious environmental shifts and narratives that communicate urgency. Films such as Avatar, Wall-E, 2012, and The Day After Tomorrow might actually help in this regard for they attempt to make spectacular slow catastrophes like deforestation, resource consumption and waste disposal, and climate change.

In this connection, it is interesting to note that during the rise of the modern environmental movement some male leaders of the movement such as John Muir were ridiculed because their interest in environmental protection was perceived as gender transgressive (Rome 2006). Reaching historically further back, Ellen Swallow Richards, one of the first scientists to use the work "oikology" to describe the study of earth systems, became known as the founder of home economics (New York Times 1895). But as an interdisciplinary scientist whose realms of investigation included the lithosphere, hydrosphere, and atmosphere, by her definition, oikology was the study of our dwelling place, the Earth. As a metaphorical direct ancestor of Noble prize-winning plant geneticist Barbara McClintock who described herself as having a "feeling for the organism," Ellen Swallow Richards was motivated by both wisdom *and* compassion. With matters of the heart belonging primarily to the female realm in Ellen Swallow Richards' time, it is not shocking that her multifaceted motivation for earth care, her form of "housekeeping," was not considered to be science. Thus, today some know Ellen Swallow Richards as the founder of the marginal discipline of home economics, not the science of ecology. Is it possible that the 21st century need to spectacularize slow violence descends not only from a paucity of imagination with regard to scales of time and an emphasis on scientific detachment, but also a need to masculinize compassion? Nixon describes creeping, convoluted and sluggishly connected changes in the earth's systems as lacking "visceral potency." The seeming need to catastrophize and harden compassion is potentially another challenge of waking up a world whose environmental and Buddhist leaders are mostly male. Might an androgynously embodied kind-minded and wise-hearted being serve as a 21st century icon to symbolize a universal effort to counter the slow violence of environmental degradation?

Because of the way we dwell in the eternal moment, I see possibly mitigating approaches to slow violence in the connected work of earth scientists and Buddhists. Sitting beside the Buddhist pillar of wisdom, geoscientists can provide images that will render vividly visible global environmental degradation over time; sitting with the pillar of ethical conduct, Buddhists can illuminate the path of heart-felt compassionate action. This view resonates with the sentiments His Holiness the Dalai Lama (2006) articulates in his book The Universe in a Single Atom: the Convergence of Science and Spirituality in which he asserts that science can contribute to the alleviation of suffering at the physical level.

During the United Nations climate conference in Copenhagen my heart and mind returned to 1990 when I worked with colleagues to map geologically a portion of the Karakoram Range in North Pakistan. During our circuitous and bumpy jeep journey following the Karakoram Highway along the Indus River and with every step leading towards our field base at Fairy Meadows (Fantori to the locals), at nearly 11,000 feet, I inhaled the scenery. Traipsing across lush alpine meadows, agog at terraced agricultural slopes, looking forward and up, I saw Nanga Parbat— "Naked Mountain"—towering at 26,660 feet, the ninth highest peak in the world. Raikot Glacier yawned beneath its north face. Around us knife-like ridges of rock, arêtes— formed where two glaciers erode parallel U-shaped valleys—separated portions of the mountain. Raikot Glacier, more ice than moraine, was healthy and frozen enough that we could walk across portions of it in search of outcrops that would give us clues to the history and rate of uplift of the Karakoram Range.

Nearly 20 years later, in June 2008 after the Chinese government shut down and then reopened entrance to Tibet, I traveled to the ancestral home of His Holiness the Dalai Lama. Driving from Lhasa and Shigatse, the second largest city in Tibet, just north of the crumpled zone where the Indian subcontinent smashes into the Asian lithospheric plate, I saw glaciers of the Himalaya once again. En route to Yamdroktso, one of Tibet's three holy lakes, I viewed but dared not approach the Kharola Glacier. Of feeble extent, this shrunken and dripping remainder of a once sturdy sheet of ice and rock manifested the slow violence exacted by human beings on the planet. We need no further data to confirm what is visibly evident, but we must behold it. Twenty-first century technology makes possible and geologists can be our kind-minded guides to the at-our-fingertips digital access to the instantaneous and high speed presentation of what in fact are slow geological changes.

The Majjhima Nikaya narrates the Nativity of the Buddha as follows (Nanamoli 1995): "…As soon as he was born, the Bodhisattva planted his feet flat on the ground and, turning towards the North, took seven steps, sheltered by a white parasol. He contemplated the regions all around and said, with the voice of a bull: 'I am the highest in the world, I am the best in the world, I am the eldest in the world; this is my last birth: for me, henceforth, there will never be another existence.' I wonder, as he stood transcending space and time, how did the Buddha view the white parasol of glacial ice in this place geologists call the roof of the world? With the greatest concentration of glaciers outside the north and south poles and rising at geologically rapid rates near 10 mm per year to the highest elevations on earth, we geologists call this region the Earth's

182

third pole. Its rapid rise affects atmospheric circulation, the breath in and breath out of our planet. With planetary warming and glacial melting, how with wisdom and compassion would the Buddha have regarded the leaking storehouses of water for the great braided rivers of the world?

By paying attention only to the knowledge of earth scientists, albeit visual and time-transcendent, we have ended up with approaches to climate change and other types of slow violence that lack heart. Skillful approaches to our environmental woes will be characterized by both wisdom and compassion; they will need to be kind-minded and wise-hearted. Only with head and heart, with compassion and wisdom together, will humanity be able to respond to slow violence that takes its toll on all living beings on the planet.

As I write this, world leaders have acknowledged that the Copenhagen conference did not result in signed agreements between countries because the distance is too great between the expectations of developing and developed countries on concessions related to greenhouse gas production. Haiti lies devastated; images from there bring to mind the mental pictures I constructed while reading Octavia Butler's (2000) Parable of the Sower. At present count, 150,000 people have died and countless others suffer catastrophic injuries, not so much because of an earthquake, but because they were crushed by collapsing buildings that were constructed on unstable ground in an impoverished country. That nations are so detached from one another when it comes to the ethical conduct of right speech, right action, and right livelihood is a form of violence in itself. What we saw in Copenhagen and in what remains of Haiti is a manifestation of violence—structural, slow, symbolic.

I sit in evidence of the eternal present. On this geologically young island in the Atlantic Ocean, all around me I see evidence of the Great Time. From hill to shore I step though time, across sandstone and clay, through fossilized colonies of coralline organisms that once lived underwater. Slopes down which I walk trace the contours of ancient sea shorelines. Coastal terraces in the coral tell me that where now I step, once was water. As I swim in the tropical blue water I contemplate the living descendants from the older reef struggling to survive in the warming sea, standing vulnerable yet vigilant before sea waves and storm surges. What Eliade once called the cyclic character of cosmic Time is here on full display. Here we have the repetition to infinity of the same phenomenon: creation - destruction - new creation: events, moments, movements.

In geological terms we live in the Holocene Epoch, the Recent, which began with the ending of the last (Pleistocene) ice age. Some have suggested that we have moved into another epoch, one that would aptly be called the Anthropocene for the dominance of human effects on the planet. In the Indian doctrine of cosmic cycles, we find ourselves in the *Kali yuga*, the fourth and last of a complete cycle (*mahayuga*) of cosmic cycles of periodic creations and destructions of the Universe. According to Eliade, in *Kali yuga*, humans and society reach the extreme point of disintegration. The 21st century portion of the Anthropocene supplies too many examples of this extreme disintegration: Hurricane Katrina, Indian Ocean tsunami, Port-au-Prince earthquake. Unfortunately they have supplied new chances for us to start from scratch in our approaches to the effects of slow violence. Whether we call this new era *Kali yuga* or the Anthropocene, may we begin to extricate ourselves from the cycles of slow violence through a green practice path based both in wisdom and compassion.

Acknowledgements

I appreciate deeply encouragement and comments on this essay offered by Rick Jarow, Joe Nevins, Judy Nichols, Rob Nixon, Michael Robertson, Ronald A. Sharp, Virginia Ashby Sharpe, and Meg Stewart.

References

Bodhi, B., 2002. The Connected Discourses of the Buddha: A Translation of the Samyutta Nikaya. Wisdom Publications, Somerville, MA.

Butler, O., 2000. Parable of the Sower. Grand Central Publishing, New York.

Carson, R., L. Lear and E. O. Wilson 2002. Silent Spring. Houghton Mifflin, Boston, MA.

Dogen, E. and K. Tanahashi (ed.), 1995. Moon in a Dewdrop: Writings of Zen Master Dogen. North Point Press, San Francisco.

Eliade, M., 1952. Images and Symbols: Studies in Religious Symbolism. Princeton University Press, Princeton, NJ.

Galtung, J., 1969. Violence, peace, and peace research. Journal of Peace Research, v. 6, pp. 167-191.

Galtung, J., 1990. Cultural violence. Journal of Peace Research, v. 27, pp. 291-305.

Ghosh, A., 2005. The Hungry Tide. Mariner Books, New York.

His Holiness the Dalai Lama, 2006. The Universe in a Single Atom: The Convergence of Science and Spirituality. Three Rivers Press, New York.

Kaza, S., 2008. Mindfully Green. Shambhala Publications, Boston, MA.

Nanamoli, B., 1995. The Middle Length Discourses of the Buddha: A Translation of the Majjhima Nikaya (Teachings of the Buddha). Wisdom Publications, Somerville, MA.

Neruda, P. and A. Reid (trans.), 2001. Keeping Quiet. *In* Extravagaria: A Bilingual Edition, Farrar, Strauss and Giroux, New York.

Nevins, J., 2002. (Mis)representing East Timor's past: Structural-symbolic violence, international law, and the institutionalization of injustice. The Journal of Human Rights, v. 1, pp. 523-540.

New York Times, 1895. Oikology a new science. New York Times, 21 July 1895.

Nixon, R., 2007-2007. Slow violence, gender and environmentalism of the poor. Journal of Commonwealth and Postcolonial Studies, v. 13-14, pp. 14-37.

Rahula, W., 1959. What the Buddha Taught. Grove Press, New York.

Rome, A., 2006. "Political hermaphrodites": Gender and environmental reform in progressive America. Environmental History, v. 11, pp. 440-463.

Smith, I. F., 1956. Where are the enduring resources? Claremont Quarterly, v. 4, pp. 54-55.

Smith, I. F., 1971. Letter to Enid H. Douglass. Claremont Graduate School Oral History Project, Interview Transcript, pp. 53-54. IFS Papers, Denison Library, Scripps College, Claremont, CA.

Stanley, J., D. R. Loy, and G. Dorje, 2009. A Buddhist Response to the Climate Emergency. Wisdom Publications, Somerville, MA.

Chapter 10: Geomimicry for Social and Environmental Justice

Marcia Bjørnerud

Summary

Since the time of the industrial revolution, Western technology has been based on an implicit view of Nature as an adversary to be defied, circumvented, or hoodwinked. After two hundred years of believing that we have outsmarted natural laws, we face the planetary-scale consequences of this adolescent attitude toward the Earth. Environmental malefactions and social injustices are arguably both rooted in a distorted sense of the relationship between humans and the natural world. Ecological integrity and human rights are equally endangered when dogma, prejudice, greed and fear obscure our shared past and common destiny. Our collective survival now requires that we learn some humility and study Earth's own history for models of dynamic and durable systems.

Introduction

In the early nineteenth century, the young science of geology literally fueled the industrial revolution by elucidating the origins and occurrences of coal. Over the next two hundred years, geology and technology matured together, as geologists gained a deeper understanding of the processes governing the distribution of petroleum, metallic ores, groundwater and other commodities essential to an increasingly voracious global economy.

Because geology has been entangled with resource extraction from its earliest days, the science is at least partly culpable for countless cases of environmental and social irresponsibility, from hellish working conditions and employment of child laborers in mines, to denuded mountaintops, oil spills, acid rain and global warming. With such a record, geology would seem an unlikely contributor to the social justice movement. But in the future, geology may supply an entirely new kind of commodity, an accidental but potent byproduct of two centuries of geologic exploration: the wisdom—both pragmatic and philosophical— embedded in the rock record.

Keenly aware of the multivariate complexity of natural systems, geologists have been reluctant to proclaim access to the kinds of universal laws that distinguish the "pure" sciences of physics and chemistry. Any particular constellation of geography, biota and climate is a singular event in Earth's history, a unique, never-to-be-revisited point on the vector of time. But when one steps back to view the grand sweep of geologic history, certain principles and patterns do emerge. Like Tolstoy's happy families, intervals of environmental health and stability are all very much alike. Environmental crises and mass extinctions, while dissimilar in detail, also share certain grim characteristics.

The chronicles of past civilizations can provide glimpses into the kinds of practices that have led societies to flourish, or to perish, in particular times and places (Diamond 2004). But historical parallels may be of limited relevance in a time when humans have become geologic agents at the planetary scale (Zalasiewicz et al. 2008). Instead, the geologic record may be our best source for insights into the factors that contribute to long-term collective prosperity—or to decline and collapse.

In this uncertain new millennium, the world needs social and economic systems robust enough to weather impending changes in global climate and geopolitics. But the world is more deeply divided than ever by religious ideologies, political entrenchment and economic disparity, and there seems little hope of finding a common philosophy or a set of principles that all nations would adopt. Even before the catastrophic collapse of global financial markets in 2008, laissez faire economists were beginning to acknowledge that their field may be less a science than a collective mythology, and that Adam Smith's "invisible hand" is not the benevolent force economic theory has long assumed it to be (Foley 2006). Nature itself may be the only common value, and the history of the Earth the only politically neutral narrative, from which all nations may agree to take counsel.

The concept of biomimicry or biomimetics—looking to natural living forms as templates for good design — has become a potent source of new inspiration in architecture and engineering. I suggest that the idea be extended to include what might be called "geomimicry"—looking to the Earth itself as our "mentor, model and measure" (Benyus 1997) in building environmentally sound and socially just infrastructures.

An understanding of the rock record makes it clear that in the long term, we really have no choice: Earth *will* be the ultimate designer and arbiter of human destiny. An irony of modern technology is that understanding natural laws well enough to build marvelous machines has caused us to believe that we are exempt from natural laws. Geological laws in particular, which are slow to have obvious manifestations, can seemingly be circumvented without consequences. A physical miscalculation may lead swiftly and catastrophically to a train derailment or plane crash, but breaking geologic rules rarely has such immediate repercussions. Geological jurisprudence is ubiquitous but largely invisible—the very framework of our existence, and therefore transparent to us. This blindness arises from our inability to see ourselves and the landscapes we inhabit in proper temporal perspective, to recognize that the natural amenities we consider "givens"—life-sustaining soil, water, and ecosystems—are not fixed but, in fact, are fortuitous legacies of past geologic circumstance, without guarantees for the future.

Our collective ignorance of natural history could be written off as harmless except that environmental malefactions and social injustices are arguably both rooted in a distorted sense of the relationship between humans and the natural world. Ecological integrity and human rights are equally endangered when dogma, prejudice, greed and fear obscure our shared past and common destiny.

Social justice, moreover, includes intergenerational justice—the rights of future generations to a planet that is livable. This principle was articulated at least 250 years ago in the Iroquois

Gayanashagowa or "Great Law of Peace," which states that leaders should take actions only after contemplating their likely effects on "the unborn of the future Nation... whose faces are yet beneath the surface of the ground" (Fenton 1988). Recasting decisions in terms of time scales longer than a single human lifetime requires a consideration of environmental and geologic factors that are easily forgotten on a day-to-day basis. Fiscal years and congressional terms enforce a blinkered view of the future; few modern political entities are configured to make plans on the scale of decades, much less generations. The single most important contribution of geology to the environmental justice movement, then, may be to instill in politicians, CEOs, and ordinary citizens an altered perception of time.

So what can Earth teach us about building systems with longevity and durability? And how can these geocentric operating principles be translated into policies that will take root in our economic and political systems and produce measurable benefits? By respecting and obeying geological "laws," rather than trying to evade or circumvent them, we may discover untapped reserves, within ourselves, of fairness, frugality, resourcefulness and common sense.

Geomimetic Principles for Social Justice

Principle #1: Uniformity of laws

If there is any geologic "law" that geoscientists cite without hesitation, it is the Principle of Uniformitarianism: the idea that the same processes operating on Earth today also operated in the past. First articulated by Scottish "father of geology" James Hutton in 1788, this seemingly obvious thesis was a direct challenge to Biblical "catastrophist" accounts of Earth's origin and history. Conceptually, it was cognate to Newton's demonstration a century earlier that Heaven and Earth are governed by the same physical laws. In a sense, Newton and Hutton together decreed a new kind of cosmic egalitarianism, with past and present, the celestial and the mundane, formerly thought to be subject to different rules, all brought together under one regime.

It is perhaps not coincidental that Hutton's <u>Theory of the Earth</u> was published within months of the writing of the U.S. Constitution. Although the two works have deistic underpinnings, both disavow the idea of direct intervention by God in the workings of the world, whether in shaping landscapes or bestowing special rights to kings. And both documents are declarations of the uniformity of natural, "self-evident" laws across time and space. But Hutton was more Jeffersonian than Jefferson himself; while Jefferson's "uniform" application of laws included some major exceptions (slaves, women), Hutton was utterly consistent in his application of the Principle of Uniformitarianism. He understood that there are no asterisks or grandfather clauses in the geologic constitution.

This is a lesson learned painfully again and again in geologic disasters, which are ruthless levelers in every sense of the word. Cliffside mansions and ramshackle mountain villages alike will succumb to rain-soaked mud and gravity. But as the receding floodwaters of Hurricane

Katrina revealed, geologic processes are not so equitable in their effects when they interact with non-egalitarian social and political landscapes. The more gradual effects of global warming are also likely to affect disproportionately the world's poor, who already live in agriculturally marginal and environmentally vulnerable areas (Dreifus 2006). The probability that climate change will have inequitable and destabilizing consequences has even caught the attention of the U.S. Department of Defense (2010).

Occasionally, geologic catastrophes can lead to new clarity of vision and unity of purpose. In Aceh Province of northern Indonesia, for example, the cataclysmic tsunami of 2004 finally put an end to a brutal thirty-year civil war over control of oil, mineral ores, and timber. The giant wave, which claimed ten times more lives than the war, transformed former enemies into fellow survivors who were willing to negotiate a peaceful resolution. Signed just eight months after the tsunami, the Aceh Peace Agreement is still young and fragile, and its success will require continued commitment to collective stewardship of the area's natural resources. But its very existence is almost as momentous as the earthquake that triggered the catastrophe (Renner and Chafe 2006).

Still, a hard, cold fact about the tsunami lingers in the back of many geologists' minds: the death toll would likely not have been so great if an alert system equivalent to the Pacific Tsunami Warning Center (established in 1949) had existed in the Indian Ocean. (To the credit of the international geoscience community, such a system has now been implemented through the Intergovernmental Oceanographic Commission of UNESCO (2011)). Geological understanding of natural hazards and the technologies for mitigating their effects should be as universal as geological processes themselves. Geologists from countries with strong scientific infrastructures naturally view the entire Earth as their laboratory. Many of us have traveled, in the name of science, to remote islands, pristine mountain ranges or forbidding deserts in impoverished nations, enjoying privileged vistas in those places and returning with heroic tales of our exploits. But if our investigations lead to no more than professional accolades and puffed-up citation indices, then our work in these geological meccas is little more than scientific colonialism.

Fortunately, a growing number of geologists carrying out research in geologically rich but cash poor regions include "autochthonous" students and researchers in these projects, thereby helping to build local scientific expertise. Professional organizations and international agencies are helping to make scientific literature available to scholars in developing countries through initiatives like Online Access to Research in the Environment (2011). In this way, we geologists have the opportunity to foster social justice through the very practice of our science. In fact, we have an obligation to share our knowledge of Earth with all its residents, to work toward a world in which the laws of Nature are indeed applied uniformly. Hutton (and I suspect even Jefferson) would approve.

Principle #2: Loopiness

Other than a fine rain of comets, micrometeorites and the occasional dinosaur-killing asteroid, Earth has not acquired any significant new mass over the last 4 billion years. Yet the median age of surface rocks is only about 500 million years, the entire volume of the oceans is replaced every few millennia, and soils are constantly washed to sea. This means that everything on the planet is incessantly recycled. The water, air, soil, the very crust of the Earth, at vastly different rates, via evaporation, precipitation, oxidation, reduction, subduction, metamorphism, melting, crystallization, erosion, and mountain building are reused, reacted and remade. In fact, this tendency for infinite repetition and reinvention is the single most important attribute differentiating Earth from its sister planets. Not surprisingly, life on Earth also learned circular reasoning early on.

In the biosphere, commodities are by necessity used at rates commensurate with their abundance. If an essential element is very scarce, organisms have only a few options: 1) learn to live with extremely limited amounts of the commodity; 2) specialize in harvesting it from the environment; and/or 3) find ways to recycle it for repeated use. None of these is easy, but given that the alternative is extinction, organisms have come up with ingenious strategies to make a little go a long way.

Biogeochemist Tyler Volk (1998) has suggested that the frugality of biological systems can be quantified by what he calls a "cycling ratio:" the amount of a commodity that is processed within an ecosystem in a given period of time divided by the amount that exits the system in the same period of time—essentially the number of times it is traded among organisms before becoming locked up in sediments or returning to the atmosphere. A resource that is reused many times via transport between organisms (e.g., soil microbes, plants, herbivores, carnivores, scavengers, decomposers) before leaving the biotic system will have a higher cycling ratio than one that is taken in and discarded after a single use.

The values of natural cycling ratios vary over many orders of magnitude, depending on what systems and resources one is considering, but cycling ratios are always *inversely correlated with availability*. That is, the scarcer the resource, the more the biosphere recycles it. Consider two elements needed by life forms: calcium used for shells and bones, and nitrogen, essential at the cellular level. Of these, calcium is by far the more abundant in the Earth's crust. It represents between two and ten weight percent of typical igneous rocks in the near-surface environment and has a very low cycling ratio, not quite two. Nitrogen, virtually absent from rocks, is abundant in the atmosphere, but not in a form accessible to any organisms except for a few experts: highly specialized nitrogen-fixing microbes in the soil and the sea. Volk estimates the global biological cycling ratio of nitrogen to be more than 1000—that is, the average atom of nitrogen is exchanged between organisms 1000 times before exiting the biosphere.

Calcium is akin to a disposable coffee cup, used once and sent to the landfill (except that in Nature even that which is discarded by the biosphere finds "utility" in the inorganic realm). I can't think of any manufactured goods that have cycling ratios even close to that of nitrogen even though many commodities we use are far less abundant. In Western economies, most goods have cycling ratios of one, or two, *if* they make it through one recycling stage. Too often, however, recycling, or even repair, is not possible because goods—especially computers, cell phones, and other electronic gadgets—are not designed to be dismantled, and parts can neither be reused nor replaced. Much of our toxic high-tech trash is shipped to developing countries where workplace safety and environmental regulations are minimal or nonexistent. This is the dark side of our beloved, brightly-lit screens.

We used to get much higher cycling ratio scores. In colonial and frontier America, right through to the 1940s, to be frugal and resourceful was a necessity, a virtue and even a patriotic duty. As historian Susan Strasser (1999) writes in Waste and Want: A Social History of Trash, "everyone was a bricoleur" (p. 22). (A bricoleur is an anthropological term for a "jack-of-all-trades," who uses few, non-specialized tools for a wide variety of purposes. The rough opposite of a bricoleur is an engineer.) Virtually all commodities had multi-stage life cycles. After being worn hard for years, mended and handed down from sibling to sibling, clothes became patchwork quilts, then rag rugs, then draft stoppers around windows. Metals were melted and shaped and remelted and reshaped at smithies. Furniture and dishes were passed down from one generation to the next.

Such frugality is still an economic imperative in many parts of the world, but in the U.S., consumption has become strangely entangled with the idea of good citizenship. In fact, the word "consumer" is now more or less a synonym for "citizen" and that doesn't seem to bother anyone. During World War II, profligate consumption was considered almost treasonous. After the terrorist attacks of 2001, we were told to go out and shop.

To be fair, there are some sectors in the modern economy in which cycling ratios reach two or better. Municipal recycling programs for paper, aluminum, steel and some plastics are well established in many communities. And thrift stores have had a small renaissance in these hard economic times. "Deconstructing" rather than demolishing houses is also becoming more common and even fashionable, allowing vintage building materials to be reused in new homes and businesses. And most road construction projects now incorporate recycled asphalt and concrete. But we still fall far short of natural cycling ratio values, even for commodities whose absolute abundance is much smaller than that of nitrogen for the biosphere.

In their book Cradle to Cradle, environmental economists William McDonough and Michael Braungart (2002) describe their vision of an economy in which most industrial processes are "closed loops" and cycling ratios could achieve double digits. We can emulate the loopiness of natural biogeochemical cycles by making disposal more expensive than recycling at every step in the industrial food chain. In Nature, the efficient recycling of nutrients provides "livelihoods" for organisms in myriad ecosystem niches, creating vibrant natural economies with full employment.

Closed-loop industrial "ecosystems" could rejuvenate urban centers where manufacturing jobs have been lost owing to economic policies and trade agreements that favor unnaturally cheap, throwaway goods.

Principle #3: Nestedness

One of the biggest challenges in understanding natural systems is finding ways to measure phenomena that occur over an immensely wide range of scales. For example, how can one determine the true length of the Mississippi River system? Each tributary stream is fed by creeks, fed in turn by rivulets of progressively smaller size. At what point do you stop counting? The tiniest rills seem trivial individually, but because they are so numerous, contribute significantly to the whole. Systems like this are examples of fractals—entities that defy traditional Euclidean geometry because they have non-integer, or fractional dimensions. A branching river system, for example, is more than a one-dimensional line, but doesn't quite fill all of a two-dimensional plane, so it has a fractal dimension of somewhere between one and two. Fractal systems are ubiquitous in Nature: Weather phenomena, mountain ranges, ecosystems—all are characterized by this kind of iterative "nestedness," with each level in the hierarchy enclosing a smaller but equally complex microcosm. Mother Earth might more appropriately be thought of as Matryoshka Earth.

If simply describing the anatomy of this sort of system is difficult, understanding the physiology—the dynamics of interactions within and between tiers—is boggling. Ecologists have known for some time that healthy ecosystems—productive and resilient ones—tend to have high biodiversity. But a new generation of ecosystem studies that combines field, laboratory and theoretical approaches is finding that it is not merely the *number* of species but the highly structured "architecture" of ecosystem networks that is responsible for their robust health (Bastolla et al. 2009). In fact, a high degree of nestedness—in which "specialist" species (like insect pollinators) interact with only a limited number of generalist species from the next hierarchical level (e.g., certain plants)—is what *fosters* biodiversity by reducing direct competition between coexisting species with similar survival strategies, just as small creeks in different watersheds don't compete for rain. Yet the structure as a whole is energized—and stabilized—by the tangle of mutualistic interactions that link all levels and components.

These principles can obviously help to inform the design of effective wildlife conservation policies, but they can also be applied to stewardship of other natural resources. Water resources in particular can be protected better if management policies are designed to work in harmony with nested natural systems. An excellent example is Wisconsin's enlightened new groundwater law (WI Act 310-03). The legislation stemmed from a proposal by the Perrier Company in the late 1990s to install high-volume wells in east-central Wisconsin for bottled water. Concerned about the potential effects on the regional water table and wetlands, an unlikely consortium of farmers, environmental groups, fishing enthusiasts, local governing boards, and state legislators turned to hydrogeologists in the Wisconsin Geological and Natural History Survey for help in drafting a bill that prevents installation of new wells that would withdraw water at rates incommensurate with local natural recharge. Rather than making a procrustean "one-size-fits-all" decree, the bill calls for watershed-specific hydrologic information to establish maximum permissible groundwater extraction rates. For example, a high-volume well that may devastate an upland ecosystem may be sustainable near the mouth of a major river entering one of the Great Lakes. The legislation is groundbreaking in its requirement that the best available scientific data be used in the regulatory process and that decisions be based on the intrinsic scale of individual catchment basins.

The nestedness of natural ecosystems also suggests how we might build more robust, inclusive economic networks. Clear signs that we are not economically well-nested include the immense and steadily growing gap between the most affluent and most impoverished; and business sectors dominated by a few outsized players that are "too big to fail." Such lop-sized distributions are not only unjust but fundamentally unstable.

After decades of federal agricultural policies that rewarded larger and larger farming operations, the grass-roots local and organic food movement has fought back and renewed the viability of small farms. According to a recent USDA report (Key and Roberts 2007), the number of very small commercial farms—those under 50 acres—has risen by about 200,000 or 16% in the last 20 years, faster than the percent increase in the number of very large farms (those with more than 1000 acres). That sounds like good news. But during the same period, the number of medium-sized farms declined by 17%, losing ground (literally) to the largest ones, which still account for a hugely disproportionate fraction of all agricultural land. And the largest and smallest farms essentially operate in parallel, non-intersecting universes— agribusiness on the one hand and farmers' markets on the other. This missing-middle farm economy looks unnatural even to agricultural economists, who recognize it as a sickly, low-diversity ecosystem. And as Michael Pollan and other food writers have argued, agribusiness is making us sick too, with its addiction to antibiotics and its soul sold to corn syrup. Making our agricultural system fractal again, by restoring the viability of middle-sized farms, is part of the cure.

Capitalist countries smugly interpreted the fall of the Berlin Wall in 1989 as an affirmation of the "naturalness" and inevitability of the free market, but when the global economy collapsed after twenty frenzied years of free trade, unfettered capitalism could no longer be defended as inherently more natural or stable than centrally controlled systems. Communist and capitalist regimes have now both witnessed the dangers of concentrating power and wealth in top-heavy structures. By embracing and emulating the byzantine, nested architecture of Nature, we can rebuild and strengthen crumbling economic and social infrastructures.

Principle # 4: Microcracy

Gargantuan creatures and enormous calamities dominate popular depictions of the geologic past. Tyrannosaurs, woolly mammoths, mega-volcanoes and giant asteroids make for better television than do cyanobacteria and imperceptibly slow changes in the land. But the fact is, Earth's most dramatic topography is sculpted by raindrops and snowflakes; and while hulking monsters have come and gone, micro-organisms have survived—and in fact been in charge the whole time.

This is microcracy: Rule by the tiny.

The modern science of geology actually began with recognition of the might of the minute—the realization that Earth's surface was shaped not by Biblical cataclysms, but rather by apparently inconsequential, incremental processes integrated over time and space. This principle was also essential for Darwin's great epiphany that the subtlest variations in the traits of organisms have given flower to the immense diversity of life.

Although the inexorable workings of natural selection have led to some truly colossal beasts, it is in no sense hyperbole to say that microbes—bacteria, algae, protozoa and fungi—govern the biosphere. They are critical to closing nutrient cycling loops by breaking down organic matter. They populate every conceivable environment on the planet and mediate the chemistry of soils, the oceans and the atmosphere. They even eat rapaciously away at rocks; acids produced by micro-organisms accelerate chemical dissolution of rocks at Earth's surface by many orders of magnitude.

In our own bodies, there are more "guest" bacterial cells than "native" animal cells. Most of these bacteria are benign and many are essential to digestion and healthy skin. But of course there are also microbial villains. If anyone is skeptical that we live in a microcracy, consider that few modern humans live in fear of being eaten by large carnivores, but a virus can put the entire world on high alert. As fearsome enemies or powerful allies, micro-organisms teach us that tiny is not the equivalent of trivial.

This is the design principle behind microfinance programs like Grameen (or "Village") Bank, founded by 2006 Nobel Peace Prize winner Muhammad Yunus. His revolutionary idea was to change the world from the bottom up by establishing a system for making tiny loans—most less than 200 U.S. dollars—to entrepreneurs in the poorest places in the world. The loans are issued at low or no interest, on an honor system, to applicants who have an idea for a small business—a dairy, a weaving cooperative—that can help lift them out of poverty. Loan officers meet frequently with clients and develop long-term relationships with them. To date, Grameen Bank has made more than 8 million microloans, about 70% of which have led to successful, self-sustaining small businesses. An astonishing 95% of the loans are repaid. Yunus attributes this to the sense of "moral responsibility" that emerges when business transactions are personal. In 2009 Yunus was awarded the Presidential Medal of Freedom by President Obama. When a reporter asked how his bank, which lends to the poorest of the poor, could be so successful, at a time when powerful banks in the U.S. are failing, he replied that while they were investing in fantasies, he was investing in goats (Ahuja 2009).

The benefits of Grameen loans spread outward through the communities—contagious like laughter. Without any such means to break free of poverty, poor people and communities often become victims of other infectious agents: parasitic loan sharks, pawnbrokers and drug-lords who insinuate themselves like mildew into more and more lives.

Although microcracy is obvious and ubiquitous in the natural world, there is something deeply counterintuitive about it, perhaps because evolution has programmed us to be more wary of large, fast-moving carnivores than slow, incremental threats to our existence. Microcracy is in essence the principle underlying democracy, but even in well-established republics, we struggle not to succumb to skepticism about whether our participation matters. We too often interpret the egalitarian dissemination of power as individual helplessness, despairing that our solitary actions, or lone votes, will have no effect on the environment, or the outcome of the general election.

We forget that the potency of microcracies, whether bacterial or political, arises not simply from the sheer numbers of individuals, but from the networks of connectivity within them. Certain strains of bacteria and even some insects can coordinate their collective behavior via "quorum sensing"—biochemical signaling that allows a decentralized population to assess its own size and respond appropriately to its environment. Depending on the species and the setting, this can lead either to healthy guts and thriving ecosystems—or to cholera epidemics and plagues of locusts.

The Internet is the most obvious anthropological illustration of the potency and peril of a highly interconnected microcracy. It has empowered millions of ordinary people by providing access to libraries of information and connections to virtual communities around the globe. But it has also spawned new viral subcultures of hate, crime and exploitation.

In massively parallel computation and "crowd-sourcing" projects, science has begun to emulate microbial systems by enlisting legions of participants to take on tasks that would otherwise be impossible to complete. Examples of successful science by the masses include NASA's "Clickworkers" project in which more than 85,000 people examined Mars images to count and classify craters (Kanefsky et al. 2001), and the "FoldIt" videogame project in which the 57,000 players who helped solve geometric puzzles about how protein molecules fold in three dimensions were named (collectively) as co-authors of a paper published in Nature (Cooper et al. 2010) Similar "bottom-up" scientific efforts could be launched for projects with societal benefit—e.g., natural hazard mapping and mitigation through community remote sensing (National Academies of Science 2010).

Microcracy, like democracy, does not guarantee good outcomes; in both systems, small changes can become magnified and lead to either prosperity or pathology. The key is to identify the "good bacteria" and foster that live and active culture.

Principle #5: Historicity and causality

There is an implicit hierarchy in the sciences that places the "pure," timeless, laboratory-based disciplines, especially physics and chemistry, above those sullied by the messiness of history—biology and certainly geology. The attraction of research that promises exact and unambiguous solutions is understandable. But in an untidy world where the idiosyncrasies of history generally invalidate the assumptions behind such elegant solutions, blind faith in "pure" science is Panglossian at best—and Pandoran at worst. Denying the importance of history leads one to forget the complex webs of causality that define natural and human systems. Many environmental and public health problems—e.g., the eutrophication of water bodies as a result of fertilizer runoff, or the pervasive presence of lead in inner city soils—are the result of simplistic (but pure!) understanding of complex systems and willful disinterest in histories or consequences. Want a smoother-running engine? Leaded gas will do the trick Yearn for a greener lawn? There's nothing better than lots of nitrogen and phosphorous. What happens next is "beyond the scope" of the pure sciences, or as satirist Tom Lehrer put it "'Once the rockets go up, who cares where they come down? That's not my department,' says Werner von Braun."

It is time to rethink the scientific hierarchy and recognize that the time-entangled sciences are just as fundamental as—and arguably more important from the standpoint of human well-being—than those that *a priori* exclude the particulars of history. We need to embrace rather than deny the "historicity" that makes natural systems so rich—and so incredibly complicated. Physicist-turned-environmental scientist John Harte (2002) has argued eloquently for a new kind of science that merges Newtonian thinking (idealized, ethereal, timeless) with Darwinian (organic, earthy, evolutionary). Historian Peter Turchin (2008) goes further, suggesting that the field of history transform itself into a quantitative science—modeled after geology and evolutionary biology—that seeks patterns of causality in the vigor or decline of societies, while also respecting the singularity of any moment in time.

In the natural world, history is destiny, and the consequences of actions (or inactions) are consistent and immediate. Natural selection will deliver a swift and unambiguous lesson—extinction—to any species that doesn't "understand" its environment. Children of our own species learn how the world works by observing patterns of cause and effect and gradually internalizing physical principals and social mores. Those from homes where discipline is absent or inconsistent may never develop a sense of personal responsibility or capacity for self-control. In societies too, a sense of individual accountability can be eroded by political and economic systems whose opacity and anonymity disempower people by obscuring causal connections between actions and their results. Globalization of the world economy has distanced us from the sources of the commodities we use. Digital technologies subtly undermine our tactile, instinctual connection to the physical world by making actions "virtual" and by reducing decisions, small and large, ethical or immoral, to mouse clicks. Modern infrastructures that make life easier, safer, and sweeter smelling—food distribution networks, electrical grids, sewage systems—also act as physical euphemisms that mask the dirty consequences of our daily habits and allow us to go about our business with unnaturally clean consciences. With environmental cause and effect as disconnected as they are today, we are at risk of becoming a "Dorian Gray" society in which the ugly manifestations of our lifestyles are stored out of sight—typically transferred to less affluent communities and future generations—while we carry on with the free trade bacchanal.

We need a better sense of temporal proportionality: to consider carefully whether the benefit we derive from extracting and consuming a commodity is commensurate with the time it will take to heal any environmental damage done over its life cycle. This is one metric for intergenerational justice. It seems quite obvious, for example, that toilet paper should not be made from old-growth forests. Unfortunately, many other products provide short-term benefits to a fraction of the current population, but will have far-reaching environmental consequences for much larger numbers of people over time. Atmospheric carbon dioxide from combustion of fossil fuels has a residence time on the order of 100 years, longer than the lifespans of those who consider consumption of cheap gasoline a birthright. Other types of waste will persist far longer. The mining of gold, which in even the richest deposits is measured in ounces per ton of host rock, often leaves behind cyanide contamination and topographic derangement that will persist for centuries (Perlez 2005). In exchanging gold wedding bands, few couples realize that they are inadvertently creating a legacy of cyanide contamination that may outlast even the most successful marriage. Radioactive waste, with its own internal clock, forces us to think of the choices we are making for the unborn. Even Congress (whose vision rarely extends beyond a six-year senatorial term) mandated that the nuclear waste repository at Yucca Mountain be designed to last at least 10,000 years, longer than all of recorded history. While this at least shows an inclination to think of future generations, it reveals a certain naïveté about the durability of human structures over geologic time. By failing to understand time, we are stealing the future from those who will follow us.

Conversely, some of the most tragic cases of social and environmental injustice involve theft of history. The Gila River Indian Community in Arizona, once self-sufficient in agriculture, slowly withered as water from the river was redirected to ranches and corporate farms upstream. Farming became impossible, public health declined as government food rations replaced locally grown produce, and the fabric of the community was weakened by poverty and anomie. Although the Pima and Maricopa people continued to live on their traditional lands, their connection to it and to their own history had been eroded. In 2008, a Federal settlement began to bring flowing water back to the Gila River Reservation (Archibold 2008). The hope is that in addition to irrigating corn, beans and squash, the water will renew cultural relationships with the land.

The restoration of water rights to the people of the Gila River is a rare, and long overdue, example of public policy consonant with Aldo Leopold's (1948) simple measure for ethical behavior in A Sand County Almanac: "A thing is right when it tends to preserve the integrity,

197

stability, and beauty of the biotic community. It is wrong when it tends otherwise.... All ethics so far evolved rest upon a single premise: that the individual is a member of a community of interdependent parts. The Land Ethic simply enlarges the boundaries of the community to include soils, water, plants and animals, or collectively: The Land."

Many societies have arrived at this same understanding. In pre-colonial Hawaii, the principle of *kapu* set strict limitations on fishing to preserve spawning populations. Based on many generations of observation, the self-restraint of *kapu* was rooted in an understanding of the community's history and a sense of responsibility to its future. The Welsh have a specific word, *cynefin*, for this sense of a shared past, acquired when people have long dwelled in a particular place. A feeling of cynefin is arguably essential to ethical behavior, both environmental and social. If the history of a place is anonymous, unnamed and unknown, its legacy seems irrelevant and disposable. But if one knows the labyrinthine narratives of the past—the deep evolutionary roots of the modern biosphere, or the multilayered human stories of one's own city—that history becomes an integral part of the present and a guide for the future, a reference collection of best practice principles and worst case scenarios.

One of the most insidious dangers of "scientific creationism" and "intelligent design" is their blatant negation of history, a dumbing down of a sacred story, akin to censoring all but the last few words in a great literary masterpiece. When we deny who we are and the improbable story of how we came to be, we cannot see the obvious: that we are all Earthlings, with a shared geologic past and future.

Conclusion

Environmental degradation and environmental injustice both stem from a limited and ultimately self-destructive understanding of our relationship with the Earth. Like small children, we humans have difficulty imagining that the world existed before us——and does not exist solely for us. Like recalcitrant adolescents, we have spent much of our time asserting our independence from Earth. Like unscrupulous financiers, we have been inordinately pleased with our cleverness in gaming the system. Perhaps at last, we have reached a point where we can see ourselves in proper perspective, as neither the first nor last denizens of the planet, but as exceptionally destructive interim tenants who need to learn some self-control or live in squalor and conflict.

It has been proposed recently that we have entered a new geologic epoch: the Anthropocene, in which humans have become geologic agents of the first order (Zalasiewicz et al. 2008). We move around as much sediment as all the world's rivers combined; our carbon dioxide emissions far outstrip those from volcanoes, and we are pruning the biosphere as fast as some mass extinction events of the geologic past. But we should have no illusions about dominion over the planet; the fact that we are now influencing Earth systems on a global scale does not by any means imply that we are in control. In contrast to our other neuroses, this is due cause for existential terror. Fortunately, we may find prudent counsel in the planet's own meticulous records, and perhaps in studying Earth law, learn to live more justly.

References

Archibold, R., 2008. Settlement returns water to Indians, giving hope for a healthier way of life. New York Times, 31 August 2008, pp. A14-15.

Ahuja, A., 2009. Grameen shows poorest of poor can be creditworthy. Associated Press, 30 August 2009.

Bastolla, U., M. Fortuna, A. Pascual-Garci, A. Ferrara, B. Luque, and J. Bascompte, 2009. The architecture of mutualistic networks minimizes competition and increases biodiversity. Nature, v. 458, pp. 1018-1021.

Benyus, J., 1997. Biomimicry: Innovation Inspired by Nature. William Morrow, New York.

Cooper, S., F. Khatib, A. Treuille, J. Barbero, J. Lee, M. Beane, A. Leaver-Fay, D. Baker, A. Popovic, and FoldIt players, 2010. Predicting protein structures with a multiplayer online game. Nature, v. 466, pp. 756-760.

Diamond, J., 2004. Collapse: How Societies Choose to Fail or Succeed. Viking, New York.

Dreifus, C., 2006. Earth science meets social science in study of disasters (an interview with geophysicist John Mutter). New York Times, 14 March 2006.

Fenton, W., 1998. The Great Law and the Longhouse: A Political History of the Iroquois Confederacy. University of Oklahoma Press, Norman, OK.

Foley, D., 2006. Adam's Fallacy: A Guide to Economic Theology. Harvard University Press, Cambridge, MA.

Harte, J., 2002. Toward a synthesis of the Newtonian and Darwinian worldviews. Physics Today, v. 55, pp. 29-34.

Intergovernmental Oceanographic Commission of UNESCO, 2011. Intergovernmental Oceanographic Commission Tsunami Programme. Available online at http://www.ioc-tsunami.org.

Kanefsky, B., N. Barlow, and V. Gulick, 2001. Can distributed volunteers accomplish massive data analysis tasks? NASA Ames, University of Central Florida. Available online at http://clickworkers.arc.nasa.gov/documents/abstract.pdf.

Key, N. and M. Roberts, 2007. Measures of trends in farm size tell different stories. Amber Waves (USDA), v. 5, pp. 36-37.

Leopold, A., 1948. A Sand County Almanac and Sketches Here and There. Oxford University Press, New York.

McDonough, W. and W. Braungart, 2002. Cradle to Cradle: Remaking the Way We Make Things. North Point Press, New York.

National Academies of Science, 2010. From Reality 2010 to Vision 2020: Translating Remotely Sensed Data to Assets, Exposure, Damage, and Losses. Roundtable Workshop, Washington, D.C. Available online at http://dels-old.nas.edu/dr/remotesensing.shtml.

Online Access to Research in the Environment, 2011. Online Access to Research in the Environment. Available online at www.oaresciences.org/en/.

Perlez, K. and K. Johnson, 2005. Behind gold's glitter: Torn lands and pointed questions. New York Times, 24 Oct 2005.

Renner, M. and Z. Chafe, 2006. Turning disasters into peacemaking opportunities. *In* State of the World 2006, Worldwatch Institute, W.W. Norton & Company, New York.

Strasser, S., 1999. Waste and Want: A Social History of Trash. Metropolitan Books, New York.

Turchin, P., 2008. Arise Cliodynamics. Nature, v. 454, pp. 34-35.

U.S. Department of Defense, 2010. Quadrennial Defense Review Report. Washington, D.C.

Volk, T., 1998. Gaia's Body: Toward a Physiology of Earth. Springer-Verlag, New York.

Zalasiewicz, J., and 20 others, 2008. Are we now living in the Anthropocene? GSA Today, v. 18, pp. 4-8.